The Ultimate Guide to
SEA GLASS

The Ultimate Guide to
SEA GLASS

Finding, Collecting, Identifying, and Using the Ocean's Most Beautiful Stones

Mary Beth Beuke

Foreword by Lisl Armstrong

Skyhorse Publishing

Table of Contents

dedication

To my creative, patient, beautiful, and hilarious children; Elise, Blaise, and Emma, I will always be your biggest fan.

To Lindsay Furber for her business encouragement and faithful friendship.

To Teresa Crecelius and Karen Wheeler for being true mermaids and helping West Coast Sea Glass with the more difficult, day-to-day work in the studio.

To Nancy and Richard LaMotte for being more than sea glass people; for being dear friends on the other coast.

To Renne Brock-Richmond for keeping color and the arts alive in my hometown and in life.

To Lisl Armstrong for camaraderie, friendship, and teaching me to keep doing my thing.

To Todd Beuke for helping to make some of the more ambitious beach hikes happen.

To Gene, Rex, Lisa, and Julie for being my dates on those far-away trips.

To the staff at Olympic Medical Cancer Center without whose help I would not have been able to tackle writing this book.

To some of my other sea glass friends; Monica, Linda M., Ben, Jamie, Jen, Terri, Sharon, Takis, and Christeena for hard work, countless gestures of support, and friendship.

To all the West Coast Sea Glass work party ladies.

foreword

Decades ago, by way of serendipity sprinkled with wanderlust, I found my first piece of sea glass on a beach in Puerto Rico. It was a cobalt blue bottle rim. The surf literally delivered it to me. It washed up within two feet from where I was standing. When I picked it up, I felt a sense of enchantment and experienced something within my being that I can only describe as a sacred sound. My life changed and I was suddenly a full-time beachcomber. I have been wandering beaches all over the world ever since. For myself, beachcombing is a lifestyle, a passion, and a constant learning.

I first met Mary Beth Beuke in person at the 2008 North American Sea Glass Festival in Lewes, Delaware. We had been corresponding online for quite some time prior to the festival. Early on during our communications, it became apparent to me that her knowledge about sea glass was as vast as the Pacific Ocean.

We have enjoyed many phone conversations, me strolling along a sugar-white sand beach in Florida staring at the turquoise Gulf of Mexico while she overlooks the stormy Salish Sea from her deck. As a life-long learner, I was very excited to have connected with her and soon started to consider her a true resource. When someone asks me a question about sea glass that I cannot answer, or if I have a question myself, Mary Beth is often the first person I ask.

There is an interconnectedness amongst those that explore the planet's diverse shorelines. Shortly after journeying into the pages of this book, I think any beach lover will recognize a kindred spirit in the author.

Within Mary Beth there exists a perfect storm of sea glass information. She has the kind of knowledge that is rare and experiential. This kind of knowledge happens when your inner being dances with the world around you. She has been exploring beaches along the Pacific coast since childhood and lives her life according to the rhythm of the oceans tides. Mary Beth is a glass expert in her own right. Her intimate knowledge of glass types, colors, textures etc. puts her on the fast track when it comes to identifying the origin of a piece of sea glass.

Her exquisite sea glass collection is a testament of her many lengthy and meandering journeys along the world's coastlines. Within these pages, she shares beautiful photographs of some of her crown sea glass jewels. A collection such as this only comes together after many years of beachcombing far and wide.

While reviewing the materials, I found myself constantly returning to favorite sections and photographs of rare beauties. A range of topics is presented here and they all come together beautifully. You will come back to this book time and time again. This book is best viewed from a cozy perch so that you can explore, dream, and journey.

—Lisl Armstrong
artist and beachcomber
Out of the Blue Sea Glass Jewelry

PART ONE

WHAT IS SEA GLASS?

introduction

It was about 5:00 a.m. on an early spring morning. The orange sunlight beamed skyward from behind the Northern Cascades. I had to greet the sunrise for myself, so I went out to the beach, which was fortunately just a few steps away from my front door.

No one in my shoreline neighborhood had trekked out ahead of me, so footprints, paw prints, and clamming boot indents were absent along the sandy parts of the beach.

I realized I had low tide all to myself.

Any good or sufficiently obsessed sea glass hunter makes sure to move their mornings and days in symmetry with the movement of the tides. The tides change and flow along all shores of the world's oceans. Each day, there are two different high tides and two different low tides. And every day's tides land at a different time during the day than the day before.

I walked quietly and casually searched for sea glass that morning.

Looking for beach treasures has been a lifelong pastime of mine. I grew up on the Oregon Coast and spent many weekends as a child running through the sand and crab grass, building sand castles, and wading in the surf.

I have kayaked and beachcombed miles of shoreline to find, by hand, these distinct pieces, each with a journey, a captivating history, or perhaps a romantic legend to reveal.

The Northern Cascade mountain range on an early spring morning as viewed from the Salish Sea at low tide.

It is my sea glass collection that has motivated me to get out and walk and cover the shores of the Pacific and other oceans. Now, as an adult with my own young family, it is not uncommon for us to schedule family excursions and vacations around the beach.

My journey has also taken me on a quest to seek historical information about these ocean-tumbled gems from the past. So, my beach life is a combination of things; it is a life-long trek along the ocean's shores but it is also an expedition of glass archaeology of sorts.

The sea glass journey is one of intrigue, enchantment, and sometimes even archaeology. The vintage glass may have once been a piece of a colored bottle, vase, or even a schooner's lantern glass, washed up after a shipwreck.

Identifying sea glass creates a challenge between history's truths and one's imagination. How far has it traveled? What hands have held it? How long has it been sojourning at sea?

This book is your invitation to set out on that journey with me as I share personal stories of some of the world's most interesting and rarest pieces. Pulling from a vast and historic collection, I will review the distinguishing factors of dozens of sea glass specimens in all colors, shapes, and surface conditions. Join me as I explain what the ocean does to sea glass and what makes it such a coveted and diminishing resource today.

what is sea glass?

Sea glass is the small, frosty pieces of history that can sometimes be found washed up along the earth's shores. The pieces represent a timeless treasure. The journey a piece takes may have begun decades or centuries before it was found. Sea glass starts out as refuse glass that was broken and then discarded into the sea, only to find its seeming resting ground in the ocean or upon the shoreline.

The piece is then awakened as powerful elements of sand, tide, water, and weather buffet the shard over time and terrain. It is transformed during the voyage; sanded, smoothed, hydrated, and finally starts its second life as a gem. After a lifetime of tumbling, the colorful jewel washes up on the shore and waits patiently to be discovered.

what's in a name?

Sea glass is sometimes called *seaglass* or *beach glass*. Sea glass can be found as large, jagged broken chunks or as tiny bits that are about the size of the head of a pin or even as small as the sand particles they are often found amongst. If it is glass and has spent the better part of its life tumbling in the sea, to me, it's sea glass.

Some collectors, however, won't call it sea glass if it does not have a certain amount of frosting. Other collectors call it sea glass if it originates from something that was tossed into the sea as recently as last week. This is clearly debatable. For the most part, though, acceptable sea glass is well frosted on most sides and shows a true journey at sea.

Later, we'll discuss in detail where the pieces originate and how they reach the shore.

Shown here is a handful of some of the most well rounded and frosted sea glass in the world. These colors also represent some of the rarest and most sought after colors, including orange, pink, red, turquoise, and blue.

what can sea glass tell us?

Most sea glass pieces do have a story to tell. Some sea glass nuggets have been on a short journey of time and distance, but most have been at sea for decades and occasionally hundreds of years. I will take the opportunity to unlock that story, piece by piece, in some of the later sections in this book.

Sea glass is being aggressively hunted all over the world by both the casual beach lover and the serious collector. Therefore, it won't be found on our shores forever. With the rapid erosion of our shorelines, many of the planet's oldest and most wonderful pieces are either lost to the depths of the ocean or are awaiting a great storm to stir them beachward.

Time in the elements, constant pounding upon a rocky shore, plus long-term hydration all play important roles in the conditioning and "frosting" of genuine sea glass.

The color of this deep, cobalt blue flat is one of the sea glass collector's most coveted colors. Though this piece is small, it is in prime condition, showing all rough edges smoothed away by time in the shores surf and sand.

where does sea glass come from?

The quintessential question that most every sea glass collector asks is "Where did my sea glass come from?" I have spent the last twelve years of my life helping people to answer that question.

I was inside the display stanchions assisting volunteer staffers Tom and Vickie with the shards at a sea glass show a few years ago. Underneath each shard was a card describing where each piece was found. The cards were turned upside down so the public could not see the information about the items; this information was hidden so the judges would not know details about the pieces before they did their jobs. The staff needed help identifying some of the entries and wanted to know more about certain pieces.

With my Maglite and magnifying glass for closer viewing in hand, I walked around the perimeter of the exhibition. Vickie said, "Mary Beth, tell us about these pieces." I methodically proceeded, pointing out pieces of interest and sharing my observations out loud.

"Oooh, let's take a closer look at that larger dark violet boulder. . . . I bet that originates from electric insulator slag pieces found in one of the Great Lakes." I did not volunteer to be a judge at this particular show, but had offered to help the team with their job of identifying the entries, so I was allowed to turn the info cards over. The one beneath the violet chunk read . . . Lake Erie, Pennsylvania!

"Hmmm and that tiny blue stem; it originally had a flat 'sign plate' on its top, used for identifying the contents in the bottle, oftentimes indicating poison or another toxic substance."

Hundreds of spectators surround the display tables to get a look at the sea glass displayed at a sea glass show.

Another mustard yellow piece of sea glass caught my eye. I don't see that color often. "This very creamy citron looks a lot like the same unique color of some glass I've collected in the Caribbean. Has anyone turned this card over?" I flipped the card over and it said Guánica, Puerto Rico (along the Caribbean).

"This clay marble is very old. It is my understanding that before marbles were made out of glass, the first ones were created from a mix of colored clay. This is likely one of the older marbles we will see today."

I looked up from my trip around the tabletops and noticed that a gathering of spectators was making their way along the same section of display, right in step with me. They were listening with interest in what I could describe about the pieces. I chuckled, feeling a bit like one of those historians from *Antiques Roadshow*, telling what I could deduce from each find. My friend suggested that a wireless microphone would help the curious spectators hear more information about the submissions.

"Everything about this sea glass piece tells me it originated from a glass float," I said as I continued around the station. I went on to share the reasons why I was so sure:

1. It was found along a Pacific Northwest beach, which is one of the locations on the planet where glass floats can still occasionally be found.
2. The color of it is the same color as 95 percent of the Japanese glass floats that are still in circulation.
3. The color is also not a common bottle color.
4. It is *filled* with bubbles. You can see them as they look like lighter-colored dots throughout the inside of the glass. Bubbles are an indicator of older, pre-1920 glass. Most glass float pieces are usually more than fifty to one hundred years old.
5. This particular piece also has a slight curvature to it that is not as tight as a bottle glass curve would be. It has a wider curve that indicates a larger ball shape, like a glass float would have.

My coworker Lindsay found what I am sure was a piece of sea glass that originated from a Japanese glass fishing float. The orb came away from its net in the Pacific and floated across to a West Coast shore and most likely was dashed upon the rocks. The piece then tumbled in the surf for years until she found the one-inch-long beauty on a rocky beach south of Seattle. She took it to our silver studio and artistically set it in a custom silver ring.

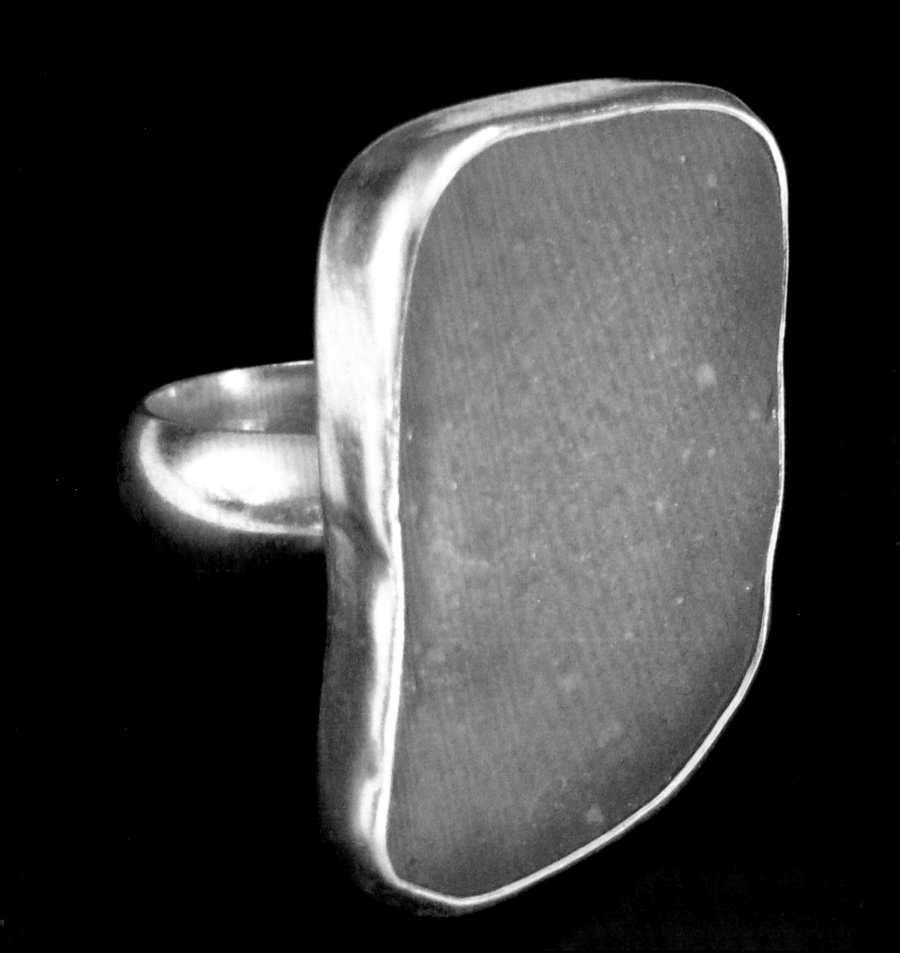

These very sharp and barely tumbled bottle pieces have been in salt water for decades. But their lack of being in an actively tumbling environment makes them quite undesirable. The sea foam green bottle lip at the top left is likely an older piece, but even the age of the shard doesn't help much with its value because it is not very well tumbled.

It is important to remember that, to each collector, their pieces are treasured no matter how common or how exceptional they are. I've had collectors come to me with garbage bags full of soiled, barely tumbled, sharp-edged, and highly common pieces who show as much pride and joy over their finds as a well-seasoned collector who's presenting a coveted, cobalt blue bottle stopper.

It is heartbreaking to see a collector who's excitedly submitted their most prized piece into a contest and come back later to view the vast color cornucopia of entries surrounding their now seemingly mundane piece. I always hold to the truth that all pieces are special, principally to the person who found them.

examples of what sea glass can tell us

This (page 17) rare, grapefruit-sized, olive green glass fishing float was brought to me for identification by a beachcombing couple. Glass floats are rarely found intact on shores these days. They were used by fisherman to keep their nets afloat; the hollow glass balls gave buoyancy to the fishing nets as the ropes were strung together and then set adrift in the ocean. Occasionally a float breaks free. Today an intact glass float is extremely rare to find. Most of the glass floats used fifty to one hundred years ago and have broken free have by now floated to a rocky shore and have broken upon the rocks.

The collectors of the dark green grapefruit float asked if I could help them to decipher what the large "NS" impressed on the button meant. They had done their research but had no luck. I explained that the letters were likely an indication of where the piece was blown and where it had originated. He asked, "But what does NS mean?"

Though I have not heard much about floats originating from Nova Scotia, I suggested this area because it has a much older history of settlement than, say, the West Coast. And olive green glass was one of the first colors of glass blown on this continent. Once I saw "older" olive green colored glass, I knew it was an older piece—probably at least as old as one hundred years.

I asked them if, perhaps, it was found in the Caribbean or along the Atlantic coast of Puerto Rico and he said yes! They were on vacation on that very island and it was there that it washed up on the beach in front of them!

This made sense to me because of the currents that run flotsam down the United States' East Coast to those shorelines. This is called the Atlantic "flow." I knew that Europe, Portugal, Spain, and Britain historically provided the main fishing fleets along the Grand Banks, and it would seem logical that there would be more glass floats from there since Nova Scotia is one of the North Atlantic's main maritime provinces where fishing trade historically used glass floats like this one.

It is known that the first glass works in Canada, called the Nova Scotia Glass Company, was up and running by the early 1880s. Interesting to me, I found that the initial manager and employer of the factory was named Mr. *Beach*.

The couple was grateful for the info and they left excited that their piece was so special and had journeyed across the sea to reach them.

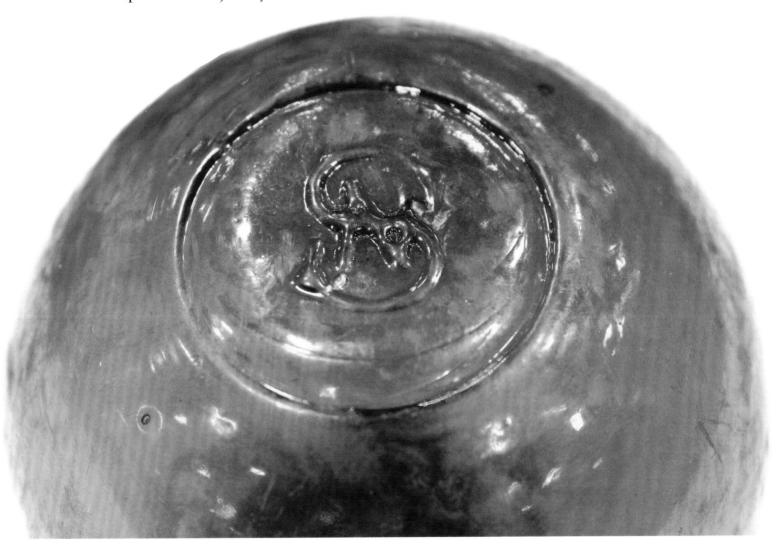

so . . . where does sea glass come from?

The primary answer is that most of it originates from refuse glass; mostly bottles and pane glass.

Sea glass comes from anything that has made its way to the sea or beach for some reason. Most countries threw their trash "away" into the sea or other bodies of water for hundreds of years. Therefore, most sea glass comes from something that was unwanted or broken. Any glassware that a person threw into the garbage was likely to make its way to the town's landfill and often, especially for coastal areas, the landfill was the beach. Though the United States now has laws to protect our shorelines from disposal, many countries still discard their garbage along the shore.

And many countries still barge their garbage out to sea, farther from shore. It is clear that it is very improbable that barged glass garbage tends to make its way back to the beach. Glass is not usually buoyant (unless it was manufactured to be a float) and therefore most glass that has been barged out to sea will sink.

Often people think our sea glass originates from items that have been dumped far out at sea and that somehow journey shoreward. This is almost never the case though. Most sea glass that has been picked up from the shore over the past fifty or more years has been discarded *near* the shoreline where it has been tumbling for decades. Currents don't carry heavy glass and deposit it along the shore, people do.

Sea glass comes from anything glass; bottles, window panes, car windshields, light bulbs, tableware, toys, marbles, figurines, and decorative items that have been tossed into the sea.

These Japanese and Korean glass fishing floats have crossed the Pacific and Bering Seas and landed on a beach far north along Alaska's west coast. They each measure about 3.5 inches in diameter. A small amount of sea glass found along the Pacific's west coast shores can be traced back to this float glass.

What about shipwrecks? It is dreamy to think about pirate plunder and long lost treasures of glassware lost at sea in a shipwreck. We dream of those artifacts being pushed ashore by a storm and waiting for the wandering beachcomber to discover them. But most shipwreck relics usually sink at or near the spot where the ship went down.

Many years ago, I discovered that my sea pottery collection had a growing number of beautiful blue and white Japanese pottery, china, and porcelain shards. I studied them closely and discovered that several pieces came from a matching dish pattern. The remarkable truth was that each piece was beachcombed from separate beaches, miles and years apart from the next.

Blue and white, Pacific Ocean pottery, porcelain, and china. Some of my pieces have been traced back to the pattern and manufacturer who created them. More about sea pottery on page 87.

Early in my professional jewelry career, I created an eight-piece, sterling silver charm style bracelet with the similar shards. What I did with it after that is a bit over the proverbial top. But I know my clientele—they are beach lovers and fantasists. This is just the thing they would go for. I wrote up the description and it sounded something like this:

"The Shipwreck Bracelet: Eight blue and white pottery pieces, once lost at sea but beachcombed from remote and rugged Pacific Ocean shores. Each has been on a lonely, lifelong journey at sea but here they seem to be made for each other. They have been reunited in this one of a kind bracelet ... blah, blah, blah."

The idea of sea glass found along our shores as being from shipwrecks is a thought that speaks to the wanderlust in each of us. But it is quite unlikely that the glass and pottery we've found has traveled across the depths of the ocean floor and tumbled up on our shores.

We now call the line the Pacific Rim line of jewelry, recognizing that much of the broken china and porcelain originates from discarded tableware, cups, and plates popular in the early 1900s across America.

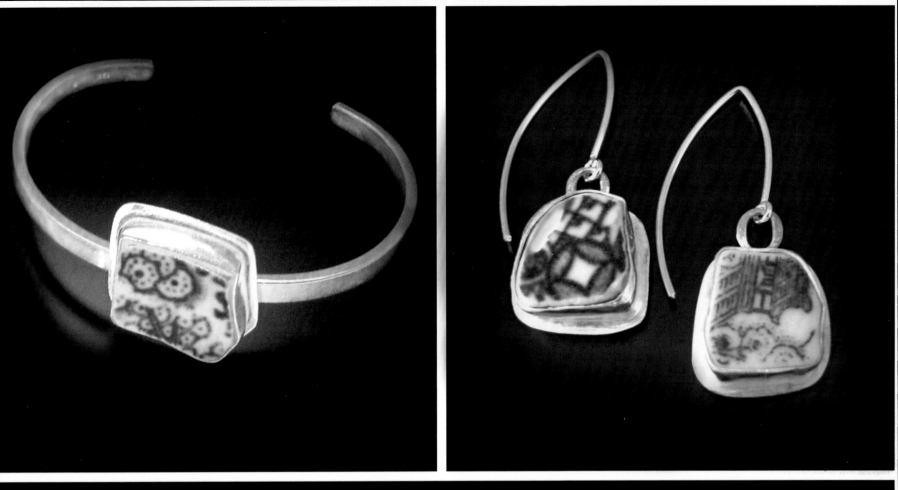

why has glass washed up on shore?

It's been called the Great Pacific Garbage Patch and it was accidentally discovered when an environmentalist named Charles Moore was sailing home from a boat race that had taken place in the Hawaiian Islands. The year was 1997, and Moore, a Californian surfer and volunteer environmentalist, was on his way back to the West Coast of the United States. He and his crew decided to make quicker time across the edge of the North Pacific Subtropical Gyre in a catamaran that Moore had built himself.

This large path of ocean is usually one that is avoided by seafarers because the warm air from the equator often pulls in the winds, which in turn create a spiraling vortex with the sea. It is almost always categorized as a high pressure area weather-wise.

According to an article in the *Telegraph* in 2009, "Several major sea currents also converge in the gyre and bring with them most of the flotsam from the Pacific coasts of Southeast Asia, North America, Canada, and Mexico. Fifty years ago nearly all that flotsam was biodegradable. These days it is 90 percent plastic."

The big island, Hawaii. Derelict fishing gear, plastic floats, soda bottles, Tupperware, toys, and more items from human dumping are washed up on an otherwise remote shore. This beach marks a southernmost point in the Hawaiian Island chain. Tons of debris from the Pacific garbage gyre breaks free and lands on these beaches.

Moore told the *Telegraph*, "It took us a week to get across and there was always some plastic thing bobbing by [. . .] Bottle caps, toothbrushes, Styrofoam cups, detergent bottles, pieces of polystyrene packaging and plastic bags. Half of it was just little chips that we couldn't identify. It wasn't a revelation so much as a gradual sinking feeling that something was terribly wrong here."

During one of my sea glass hunting trips in the Hawaiian Islands, I was lava flow and beach hopping along the very southern tip of the big island of Hawaii. It is considered the farthest southernmost point in the United States and it is there that so much debris washes up.

Though I was there on a sea glass mission, I learned so much more about tidal movement and how certain shores can grab plastic and debris. Everything imaginable had washed up along this remote stretch of beach (one that we'd been walking along for hours).

Not only does the Pacific show us proof of humans disposing plastic, glass, metals, and other material into the sea, but a simple study of the dumping practices of almost all countries that have a shoreline proves that humans throw everything into the ocean and along ocean shores and lake beaches. This is the main reason sea glass is found along the shore. At its most basic level, sea glass comes from man dumping garbage that contains glass.

Much more than glass bottles and tableware ended up along this Hawaiian shore. I have come back from beach hikes with my arms loaded with parts from cars, an entire tricycle, large chunks of porcelain from bathroom sinks, and more. Some city dumps, like Fort Bragg's in California, pushed dilapidated cars directly off the beach bluff at the town's edge.

Sometimes, one person's trash can truly be another person's treasure; some sea glass that originated from things like car windshields and bathroom shower doors make up for some of the pane glass that is found along the shores.

In the summer of 2008, I helped the Travel Channel do a show about collecting sea glass. They flew me down to San Francisco where we kayaked through the bay and searched for beaches with sea glass. The bay was a busy shipping lane. We paddled parallel to the beach as the wakes from the passing cargo vessels tossed our tiny kayaks about easily.

I was there to help tell "the story" of the pieces we unearthed. One strand of beach we discovered was a very wide stretch of rocky shoreline at the base of a bluff that was packed with weathered garbage; many broken bottles from the soft green Coca-Cola bottle days, slabs of construction wood, nails, and other discarded materials. With every step, I saw porcelain toilet parts, dark green wine bottle glass pieces, chunks of building bricks, and more.

Shipping history and marine exporting are another reason why sea glass is found along the shore. Prior to when ship garbage was treated on board (incinerated, pulped, or ground up) for discharge overboard, garbage was known to be off-loaded once the vessel reached the shore. The residual glass that has toppled over and over with the passage of time and wave action accounts for why some sea glass has washed up along shores near marine ports. We will discuss this further in the section titled "When and Where to Go."

A very flat, frosty white sea glass trapezoid likely originates from bathroom shower door or "privacy glass."

can sea glass still be found today?

The sea glass window is closing. This is what I told Margaret Zackowitz, a writer with *National Geographic* magazine while she was interviewing me for a sea glass article. But that was back in 2007. I believe it has become even scarcer since then.

Today, there is less packaging and more recycling; therefore, the pieces that were what I would call "prolific" along the shore just thirty years ago are becoming fewer and farther between. We are also seeing that the more well-known dump sites that used to have "a lot" of sea glass are availing smaller and smaller pieces.

The glass beach on the Hawaiian Island Kauai is one specific beach we are closely watching. Most pieces that remain there are so small they are just as finite as the grains of sand they are found amongst.

"I can't find sea glass to save my life," said one of West Coast Sea Glass's Facebook followers recently. He lives along the north shore of the Gulf Coast. This is significant. I have been fielding calls for almost a decade from people all over the United States—and internationally—who want to know where to find sea glass. No one who's phoned from this guy's part of the continent has ever shared with me that they've found any noteworthy concentration of historic sea glass there.

Rare colors of sea glass like blues and Depression Era pastels are becoming increasingly harder to find with each passing year.

blame it on the plastic

According to FoodProductionDaily.com, "Plastic bottles and jars are forecast to experience strongest unit growth in the European larger packaging sector to 2015 and look likely to overtake glass containers within a decade."

It is true. All over the planet, glass has been phased out, which subsequently leaves it out of our landfills and absent from our shorelines.

I have personally noticed a vast decrease in sea glass—especially in regards to the availability of certain colors—in just the past fifteen years. I know that currents and storms can change shorelines from year to year and this effect does make a difference for the beachcomber. But my colleagues and I would guesstimate that the decline is more than a 50 percent decrease on almost all beaches that we have frequented throughout our journeys.

Though I believe the sea glass window to be closing more quickly than we ever anticipated, the answer to the question posed at the beginning of this section is yes. Historical, decades old sea glass can still be found today.

why is sea glass getting harder to find?

Even the more common colors of sea glass like frosty white, amber brown, and emerald bottle green are gradually diminishing from along shores all over the world.

We've previously discussed that although sea glass can still be found today, it is getting more difficult to find with each passing day. There are several factors that point to its lessening availability.

Glass bottle production has slowed greatly since the 1970s due to the increase in plastic containers and aluminum canning that has replaced what was once mostly glass-dominated packaging techniques.

The rise in ecological awareness has caused humans to throw less refuse into the ocean or along shorelines as was done in earlier centuries. And recycling of glassware is practiced worldwide, much more today than even a decade ago.

The gradual rising of sea levels due to global warming and polar ice melt seems to be indicating that our shorelines are covered by more and more water each year.

I recently visited the shore along the mid-Atlantic coast and made a point of making my way to the beach with the purpose of touching my toes in the sand. I crossed the

boardwalk and looked down the vast beach only to witness a bulldozer moving massive piles of sand around on the beach. I had never seen such a thing before. Being from the Pacific Northwest, we rarely see vehicles on our beaches, and if we do, it's in a very few specified areas for the purpose of recreation. If I do see a vehicle on the beach, it is usually an SUV or an off-road recreational mobile. However, I've been told by several of my friends that many coastal communities do truck-in sand and bulldoze more sand onto their beaches for the purpose of slowing the effects of erosion.

If this is true, then much of the waste and sea glass that was dumped years earlier is gradually being buried deeper beneath the sand or it is being slowly covered by the ocean's gently rising tidelines.

The mass popularity of sea glass has been growing exponentially the past twenty years. Sea glass has become a rock star (excuse the pun) of sorts. When I was a child collecting sea shells, driftwood, and sea glass along the Oregon coast, I knew no one else who collected; not even my siblings.

By the mid 1990s, I knew of about four people who did. Now I rarely speak to someone who hasn't at least heard of sea glass. It's made a name for itself on the Internet, has been featured in every major publication in the United States, and now has books and TV shows dedicated to it as well as international conventions. This mass-media attention has created an entirely new collector-following for sea glass.

The *Seattle Times* ran an extensive article on sea glass in May 2009. In his excitement, the writer shared directions to an otherwise remote location that *used to* have fairly good sea glass available. The next Saturday at low tide, a fellow collector counted almost one hundred "new" collectors and tourists on the beach which, prior to the article, usually had two or three people walking it on a normal day.

Sea glass is a beautiful artifact that is running its life course. We must remind ourselves that it once was garbage and most probably will one day disappear from our lake and ocean shores.

what is lake glass?

Essentially, sea glass is any piece of glass that has been affected by the tumbling of waves, movement of tide, and submersion into sea or lake water. Many people will call glass that has been found along a lakeshore *beach glass*. To most collectors, these names are interchangeable—beach glass is sea glass and sea glass is beach glass. To the highly particular purist, *lake glass* can only also be called beach glass and cannot be called sea glass because sea glass is *only* that which has been found along a true sea or ocean.

These purists note that lakes don't really have waves and tides. This is undeniably true. But large lakes, like the United States and Canada's Great Lakes, do experience fluctuations in their shore water levels because of winds and barometric pressure changes.

Also, the Great Lakes experience true tides—high and low tides on a semi-diurnal pattern. This means twice daily the tides change. Some days this fluctuation can be close to five centimeters.

and river glass?

River glass is the name given to the historic pieces of bottles and glassware that tend to be found along river banks and river mouths.

One of the best locations to find sea glass is where a river meets the sea. That is because long ago people were known to dispose of their garbage into rivers. When the river runs down to the sea, especially after a storm, garbage can be washed to the ocean shore.

My daughter's dear friend Victoria tells the story of a summer night when she was fifteen years old and went on a long walk with her dog, Bear. The two of them crossed a bridge over a river and then they hiked down, parallel with the bank of the river until it reached the sea. The river mouth had opened up wide onto the beach. It was getting late and she decided that a quick way to save time getting home was to cross back over the river at the mouth. There was no bridge to connect them to the other side. But because the river had widened onto the beach, she figured it may be shallower there than it was upstream.

Bear has always been skittish about the water, so Victoria bravely helped him up onto her shoulders. She stepped in and, within a few paces, the water level was at her thighs. By about the center of the river, the level was clearly at her ribcage and the current was beginning to push strong against the two of them. She looked straight ahead and took another step. The water was suddenly up to her shoulders. Bear pulled and squirmed with fear. She had no choice but to turn back, forcing strong strides to find her footing toward the riverbank they had just come from. They reached the shore and Victoria found herself face to face with a muddy, waist-high embankment.

Directly in front of her, stuck right into the façade of the riverbank, was an oddly wedged glass bottleneck. Curious, she pulled at the two-inch protrusion and realized that the body of the bottle was attached and buried deep into the damp bank. She began excavating chunks of the clay wall away from the bottle, wiggling the relic free.

What she found was an old, intact gin bottle. The piece was marked GORDON'S DRY GIN and the embossing showed Gordon's famous trademark boar's head symbol on the bottom.

When she told me the story, I expressed to her that the treasure's location didn't sound so strange to me as beachcomber. I shared with her that, historically, many communities disposed of their garbage into the river and that river mouths are one of the best sites to find vintage bottles, pottery, and sea glass.

why do we love sea glass so much?

I love sea glass collecting because it gets me to the beach and, in my personal journey, it got me *back* to the beach.

It was March and springtime was coming. I had just finished an unexpected seven months of chemotherapy for a cancer that I didn't even know was inside me that many months earlier. Chemotherapy wipes you out physically. Even when one has completed several months of treatment, the medicine remains in your system and it can take months to regain energy.

This was the case with me. Even though I was finished with treatment, I still had weeks and weeks to go until my energies were back to the point where I could comfortably enjoy a brisk walk on the beach for any more than five minutes.

But I did get back to the beach. And so I started my recovery jaunts off with slow, metered steps. I would walk slowly to the edge of the crabgrass and step tenderly down onto the sand. I would breathe slowly and stop often, straighten up, and breathe slowly again. It was nice. I would quietly move forward at my own pace down the beach.

This careful movement along the shore also gave me a new enthusiasm for sea glass collecting. I took the time to observe things like the meandering string of tide line detritus winding parallel to the waves at my side. There isn't always one line marking the ocean's deposits. And sometimes it is undiscernible to see much of a line of tidal markings at all.

When you walk slowly and look closely, you will observe additional beauty along your voyage.

a passion for sea glass

We love sea glass for as many reasons as there are sea glass pieces that have been found!

For some, finding sea glass is simply an added surprise and serendipitous "extra" to their love for the shore and their occasional beach walk. They are beach lovers and finding sea glass just adds to their affection for everything oceanic or nautical. They have pieces intermittently dropped into jars or along window sills about their home, but nothing much that's gathered with intent or purpose. They consider the sea glass something that the sea has "offered to them," no strings attached; beautiful, frosty gifts from the sea.

For others, the love for sea glass connects directly to the thrill of the hunt. These collectors are a bit more zealous in their search. They purposefully walk with the subsiding tide because the descending tide offers new pieces, usually with each new day. These hunters head out early with a resolve to discover and pull together a handful or more. The shoreline calls to them with gifts to uncover and collect. Each moment and each step is fresh and comes with anticipation and expectation.

And for others, the collecting endeavor is solely about an archaeological discovery. There is history to each piece; these stones have journeyed far. Their past lives must be uncovered and discovered and explained!

A seafoam green piece was once an historic bottle stopper from Ireland. A red shard was once a distant schooner's lantern lens, lost at sea in the Atlantic. A thick, frosty-white piece originated from a decades old milk bottle bottom found along a Canadian shore. Cobalt blue pieces emerged from historic poison flasks, broken and tumbled along an Aegean island shore. A dusty, olive green "smoothie" tells of a passage along a Caribbean island beach.

For me, I believe that I love sea glass for its nostalgia and its connection to history. I'm quite sure this is the main reason why most collectors have a deep adoration of sea glass. It is fascinating to me to hold up a shard and wonder all of those key questions about a piece and then proceed to answer what I can.

I have been blessed to have had the opportunity to view hundreds of people's finds from all over the world. And I consider it an honor to help them sleuth out what can be known about a piece. That is the most satisfying role I can have within the world of sea glass—the opportunity to experience countless collector's prized pieces and then to journey with them on a quest for answers.

From the more common sugary white milk bottle shards to the elusive, centuries old Roman art glass pieces, I can truly say I've beheld tens of thousands of unique specimens.

We are enamored with sea glass because, in all actuality, each piece has been on a unique journey. If only the sea glass could talk, I'm sure we would be privy to salty tales of centuries at sea!

a search for answers

There is an elusive attraction to sea glass's age-old questions. And every piece has a different story than the next piece. From where did it originate? What was it used for and whose hands have held it? How did it make its way into the ocean and how long has it resided there? And from how far away has it traveled?

For the existential sea glass lover, beach glass is a life metaphor too. How refreshing to know that in some ways, sea glass has been on a lifecycle journey just like people experience. It originates from something that was once broken, then discarded. It tumbles and cartwheels for years, often in a briny environment, buffeted against the rocks and hard places, and is grated against the grains of time, terrain, and the ebb and flow of all that "life" offers.

On more than one occasion, sea glass enthusiasts (usually women) have shared their unfathomable adoration for sea glass in a way that transcends far beyond the intrinsic visual beauty of a piece. I've had customers profess to me that deep down they really are "a mermaid." They are in awe of my job and erroneously envision me "working" from a beach chair, sitting in the sun with an umbrella drink in my hand. They ask if they can come and work with me at my romantic job, combing the shore for the rarest finds, kayaking to deserted islands, and creating one-of–a-kind works of art. Their view of sea glass is rich and ethereal and life giving and each piece holds more than meets the eye.

Though most sea glass is refuse that's been tossed onto a beach, a few pieces really may originate from pirate plunder or shipwreck goods. Imagining this can set the lore lovers a-dreaming. And the childlike side of us kicks in and spirits us away.

One such name for beach glass given by dreamers like these is *mermaid tears*. The legend tells of the heartbroken mermaid, waiting on shore for her love who's away at sea. Once her tears touch the sand, they turn to glass.

While judging the Shard of The Year contest, I always see entries from the common to the uber rare. In the foreground, this rare orange art glass piece was found by collector Anna along Italy's Tyrrhenian Sea shore. You can see by the sticker dots (that's how we voted on any remarkable pieces) that this particular find impressed a couple of the other judges, too.

PART TWO

TYPES OF SEA GLASS

first, a bit of glass blowing history

Though naturally made glass has been in existence since the beginning of time, glass blowing didn't begin until the first century BC in Mesopotamia. Obsidian, which is a naturally formed volcanic type of glass, was used by early man as a cutting tool.

lightning and meteorite glass

Most naturally made glass involves high heat within a silica sand mixture. Sometimes called *lightning glass* or a *fulgurite*, this natural phenomenon glass is created when a lightning strike or a very hot meteorite of 3,200-plus degrees Fahrenheit hits a sandy beach or desert patch. Within about one second, sand grains fuse together and the silica melts, creating a rugged tube of glass where the lightning or space junk struck.

When I work my sea glass booth at art shows, I almost always ask the patrons if they know what sea glass is. Currently, most passers-by have a basic understanding of what sea glass is and that it originates from discarded bottles and glassware that's been tumbled by the sea. But on about three different occasions over the past fifteen years, I have had a few shoppers guess that sea glass originates from lightning glass. Lightning glass has been known to be found on sandy beaches, but unless it's been tumbled by the ocean, it is not sea glass. The most popular piece of potential meteorite glass is housed in the necklace found when King Tutankhamun's tomb was excavated in 1922. For decades, the origin of this rare piece of natural glass in the center of the ornate beetle-shaped pendant was unknown. All that was known was that it had been preserved in the three-thousand-year-old tomb and that it pre-dated man-made glass production.

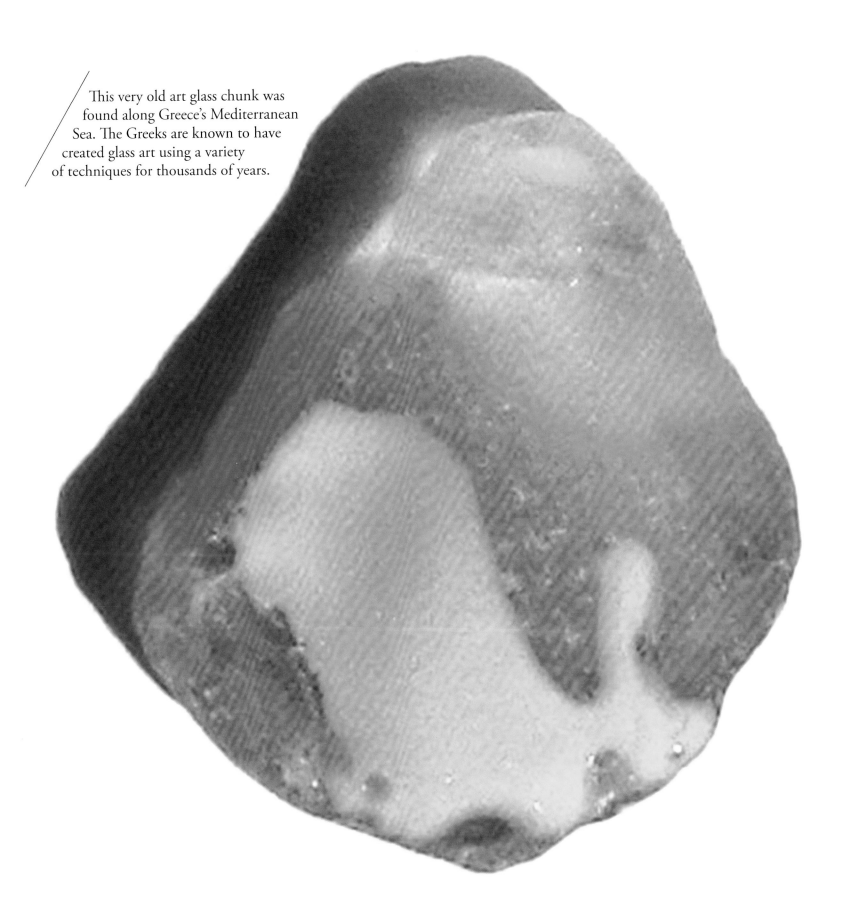

This very old art glass chunk was found along Greece's Mediterranean Sea. The Greeks are known to have created glass art using a variety of techniques for thousands of years.

So where did that first piece of yellow-green glass jewelry from the Tutankhamun dig originate? Scientists believe a meteorite broke up as it entered the Earth's atmosphere near an area called the Great Sand Sea.

A fireball with a temperature more than 3,200-plus degrees Fahrenheit would have hit the sand and melted it, making it like hot lava. As the piece cooled, a nugget of glass formed.

Man-made Glass

According to House of Glass, Inc., "the accidental discovery of how glass is created was suggested by the ancient-Roman historian Pliny (AD 23–79) as an accident by Phoenician sailors. It is possible that it was a result of shipwrecked sailors building fires for their cooking pots on blocks of soda (natron) on top of beach sand. By morning, the melted sand and soda mixture would have produced hardened glass. It was more likely that potters, from Egypt or Mesopotamia, discovered the brittle treasure independently, when firing their wares."

It is believed that man-made glass has been in existence for thousands of years. Glass creating was started by the Syrians, inhabitants of a country that is known to have the worlds longest constantly inhabited city. Glass has been used for various styles of bottles and utensils, mirrors, windows, and toys.

But it was first used for utilitarian purposes and it grew in production and creativity from there. Historians think it was first created around 3000 BC, during the Bronze Age. And Egyptian glass beads date back to about 2500 BC.

Several cultures including the Chinese, Greeks, and Romans are known to have advanced glass creation from functional pieces to more artistic forms. And Murano, Italy, is known as the home of luxury art glass making. We move west from there and follow the beautiful story of glass use and its creation in bottle manufacturing. Man-made bottles and tableware pieces account for most of the sea glass that is found today around the world.

The production of window pane glass and glass bottles became popular in England in 1557. In May 1623, Englishman Sir Robert Mansell used a "method of making glass with sea coal, pit coal or any other fuel not being timber or wood," according to the book *Early American Bottles and Flasks*. Mansell was later granted the first patent for the manufacturing of glass. The buildings in which this glass was produced were called glass works factories.

The United States opened its first glass works factory in Kensington, Philadelphia, in 1771. Needless to say, it is possible that, if they originated from American bottle products, some of the oldest sea glass pieces found along the shores of the United States can be more than 150 years old. A collector who finds more ancient pieces on US soil may very well possess pieces that were carried to the states by ship or trade vessels.

antique bottles

Notice the soft blue flask in the foreground reads Dr. C. W. Temples. This bottle in very good condition is highly rare as it held a patent medicine circa 1880.

Most sea glass found in the United States in the past fifty years originates from antique bottles. Glass manufacturing became industrialized in the United States in the early 1900s. The bulk of the glassware created was for that very booming bottling industry. Everything was bottled in glass: Coca Cola, food bottles, medicinal bottles, canning jars, milk bottles, drug bottles, pickling bottles, flasks, bleach, and other chemicals.

If you are a sea glass collector, you hold pieces of the past, mostly from bottle glass that has made a journey into the sea. Your colored pieces can tell stories about the types of bottles from which they originated. Many times, color is the main indicator, but there are also some other physical indicators that will define your piece as bottle glass rather than another type of glass.

Antique bottles and flasks, like these, most fairly small in size (under 6 inches tall), were used for everything from tonics and powders to medicines and seltzers.

As you can see, some nice blob top, bitters, and Codd bottles line the Bottle Identification table at one of our events in Pennsylvania. We came prepared for the crowds and the questions. The gentleman manning the table borrowed our bottle identification books for the entire weekend and told us he "had the time of his life" reading through bottle history and gaining even more knowledge.

bottle rim lines and threads

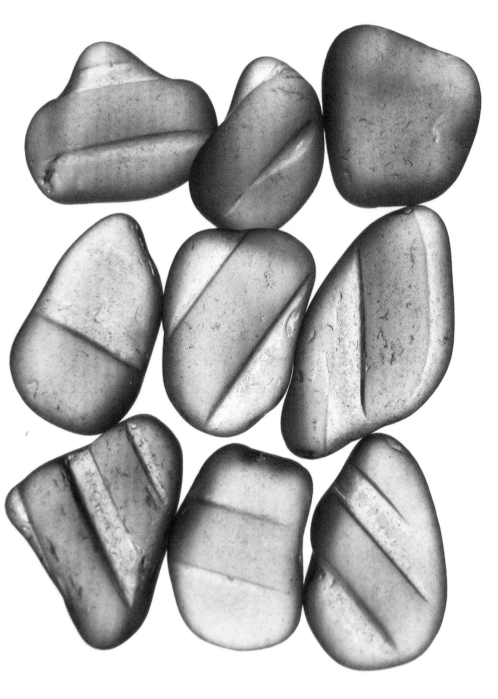

Bottle rim lines are one clear giveaway that your sea glass originates from bottle parts.

Notice the cornflower blue pieces. I know they originate most likely from pharmaceutical or medicine bottles, mainly because this color was so often used for those types of contents. Here, the lip rims, sometimes called *threads* also confirm that these pieces stem from bottle parts.

Bottle lips and rims are a more infrequent find than the other flatter parts of the bottle body because a rim or neck piece only makes a fraction of a bottle's total mass. This factor causes them to be found less frequently than other surfaces of a bottle.

These rich aqua blue shards have rim lines and threads through them, indicating they are either bottle lips or screw lines. They most commonly originate from glass canning jars or else they derive from the similarly colored electric insulator glass.

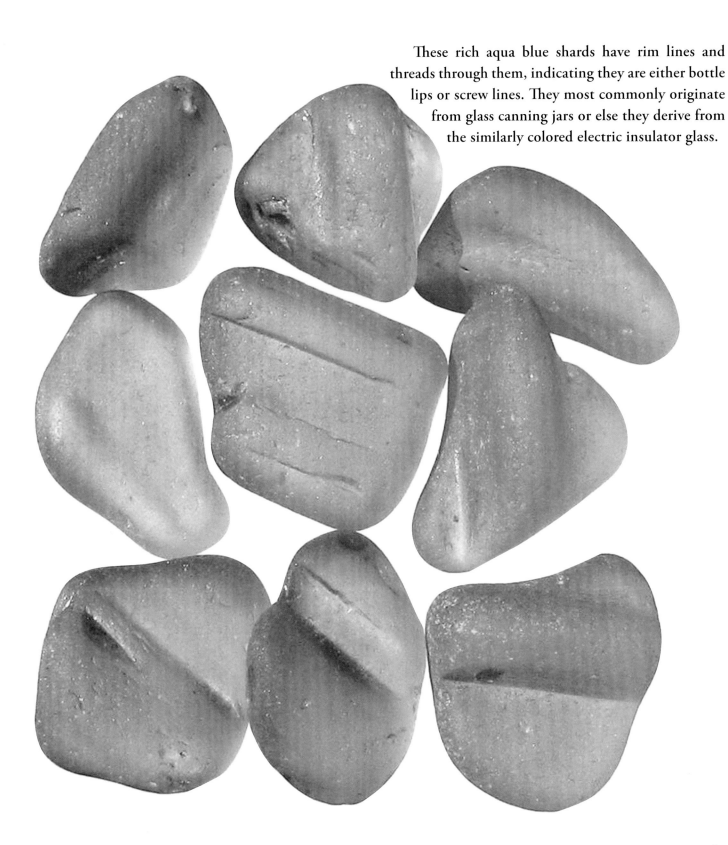

lettering

Another indicator that your sea glass is from bottle glass is if it has writing or letters on it. This lettering is usually raised and not painted or stuck on with a label and can also help to determine where your piece originated. It can also help with dating your find.

Raising letters during the molding process was an effective way to identify the bottle manufacturer and the contents in the container. Raised and recessed labeling was done before the use of easier stenciling or paper-made labeling was introduced.

Some items other than bottles did use raised and recessed lettering, such as novelty ashtrays, specialty plates, and even signal lenses. (See the Chevy taillight lens on page 156.) But, in most cases, raised lettering tells you your piece is from bottle glass.

In the late 1800s, medicine flasks had raised lettering that identified the druggist or physician. These bottles are now collector's items. I helped classify a soft blue container, similar to the one pictured at right, for a patron who stopped by the Bottle Identification table. It read: Henry Plenge, Pharmacist, Charleston, South Carolina. After researching this particular pharmacist, I can confidently date the bottle back to somewhere in the 1890s.

curves

One last indication that your sea glass is from bottle glass is if it has a curvature to it. Many bottles, though not all, are curved.

This deep lavender sea glass piece is slightly concave with a noticeable curvature, much like an early 1900s "sun amethyst" bottle glass possesses.

Keep in mind that many early bottles were also box-shaped or angular in shape. In addition, most all cups, bowls, and vases were also curved.

If your sea glass has lip and rim lines, raised or recessed lettering, or a curvature that's tighter than the curve of a soccer ball, for example, it is highly likely that it may be a piece from vintage bottle glass.

inkwells

Some of your rarer-colored sea glass may have originated from inkwells.

They saw their highest use in the early 1800s. By 1840, cut glass ink bottles were very popular and the US census lists thirty-four glass cutting factories by that time.

However, several inkwells were made in mold presses.

In 1939, the ballpoint pen was invented and inkwell manufacturing slowed greatly.

Therefore, inkwells have become a highly sought after collector's item, especially the colorful glass pieces that have unique color and decorative patterns.

Antique ink bottles line the "Bottle Identification" tables near us in the exhibition hall at our recent sea glass event. These smaller, colorful vessels were used for centuries and came in every color and design from the highly ornate to the unassuming daily-use style.

antique
tableware

For a long time, I thought it was by accident that I grew up in a home adorned with antique bottles and especially tableware; dishes, vases, goblets, and statuary.

My mother was an avid historic glass collector and was at it for so long that she had compiled an elaborate and vast collection.

After her twenty-three years working in full-time dental hygiene, my mother managed to allow her collection and avocation of antiquing to become her full-time vocation and moved herself into the world of antique dealing. She specialized in, of all things, Depression Era glass and tableware.

During her collecting years, she especially loved the soft blues, pinks, and "hobnail" serving dishes, carafes, vases, and plates. Each shelf and windowsill in our home was filled with soft colored, patterned glass from the early 1900s through the 1940s or '50s.

I recall feeling rather over saturated with what I considered at that time a rather gaudy collection. I like clean lines and a more open space and I do not consider myself a knick-knacky person.

Our colorful home was already crowded with seven children and sometimes felt even more so since we were always surrounded by pastel-colored glassware and antiques. I remember, as a child, playing Nerf football in the living room with my brothers—always when my parents were out. Almost weekly someone would bank a pass off some teetering statue or candy dish, knocking the piece to the carpet while everyone in the room froze with drawn breath. The fear was mixed with an anticipation that we hadn't dashed a priceless relic to pieces. Within a second, everyone would exhale and someone would quickly lay the treasure on its side or push it under the couch so it wouldn't happen again.

My mother became so knowledgeable and so well-known throughout Oregon as an expert that she was elected president of the Oregon Antique Dealers Association and served various roles between 1980 and 1990.

During my early twenties, I worked for my mother's antique business. I mainly frequented estate sales in tandem with her for the purpose of finding glass treasures. I also manned the antique shop one afternoon or so a week. She had a very complete library, filled with titles like *Glass Tableware, Bowls and Vases* and *Early American Glassware*. The educational books and references were spilling off and stacked atop her bookshelves.

By proximity to the industry and the glassware, I unintentionally absorbed an "antique glassware 101" education—something I would have never sought of my own direction. This accidental training paired with my years in the sea glass world have prepared me with random information that helps with the enjoyable part of identifying beach glass pieces.

identifying antique glass

Antique glass comes in such a variety of shapes, colors, and patterns that any given guidebook can feature several hundred examples of differing tableware, vases, and bowls.

If your sea glass finds derive from early American glassware, bowls, and vases, all indications point to its likely age being fifty years backward to the early eighteenth century.

The hunter who has a collection from the past two centuries will have pieces that include bottle parts, bottle stoppers, glass marbles, footed bowl parts, cut glass examples, and goblet stems.

If your pieces haven't been tumbled too smooth and some of the identifying nuances still show, it can be like glassware archaeology when a marking, bump, or scrolling swirl can be traced back to an actual glassware pattern.

My research has shown me that American glassware sea glass has been found more frequently along the United States' East Coast. I am sure this is because the population density and, therefore historic dumping, is much more concentrated along those shores. I see a larger quantity of glassware sea glass along the East Coast as well as a higher abundance of pieces that show more definable markings, which indicate the pieces stem from early 1900s glassware pieces. I directly correlate this particular difference to the fact that the rockier, more rugged coasts of the Pacific seem to tumble away many of the patterns and decorative surfaces of glassware distinguishers.

Another reason why I see more ornate sea glass glassware pieces on the East Coast is because the glass manufacturing industry in America grew out of the East. By the 1830s, factories like the Boston and Sandwich Glass Company in Massachusetts were successfully created with blown molds and presses that produced beautiful, colorful pieces.

Coincidentally, the site of the factory was built right along the beach. This was not done because sand, which is used in glass manufacturing, was readily available nearby but because the location created a convenient shipping stop and jump-off point for trade and export throughout Cape Cod and to farther shores. Some of the most memorable sea glass pieces I have had the privilege to identify have come from these factories along these very shores.

This graceful, pink piece of antique tableware originates from a Fenton art glass swan wing. It is the largest pink piece in our collection.

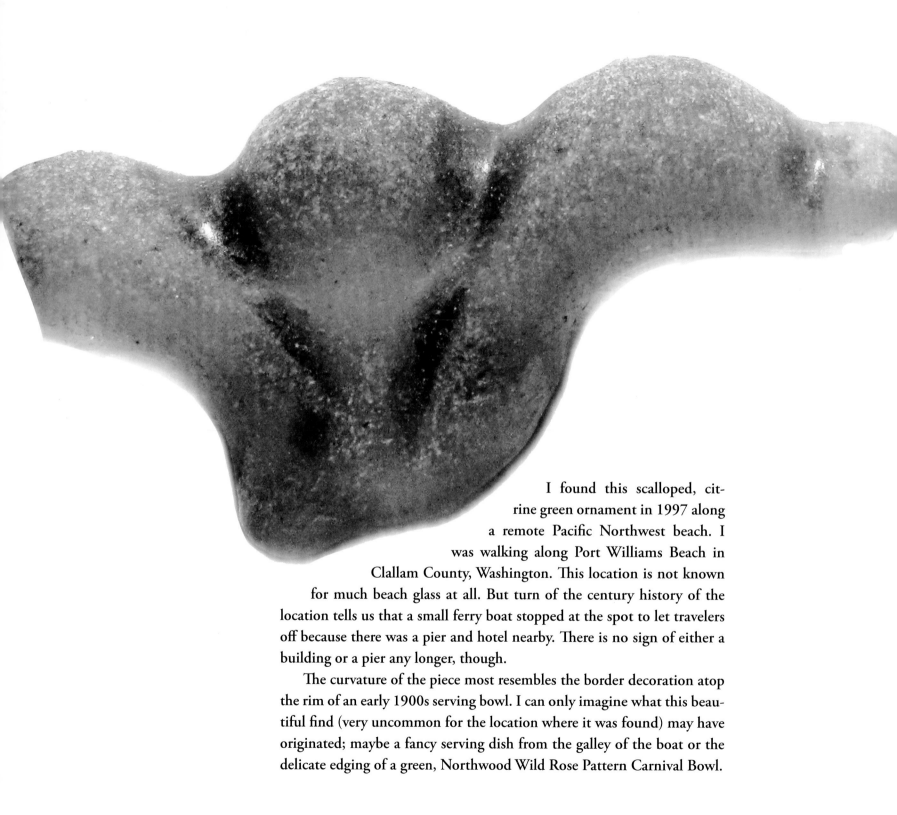

I found this scalloped, citrine green ornament in 1997 along a remote Pacific Northwest beach. I was walking along Port Williams Beach in Clallam County, Washington. This location is not known for much beach glass at all. But turn of the century history of the location tells us that a small ferry boat stopped at the spot to let travelers off because there was a pier and hotel nearby. There is no sign of either a building or a pier any longer, though.

The curvature of the piece most resembles the border decoration atop the rim of an early 1900s serving bowl. I can only imagine what this beautiful find (very uncommon for the location where it was found) may have originated; maybe a fancy serving dish from the galley of the boat or the delicate edging of a green, Northwood Wild Rose Pattern Carnival Bowl.

Deep lavender like this piece was almost never found along the West Coast unless it was found in proximity to one of the older coastal cities like San Francisco, California, or Astoria, Oregon. In addition, finding a decorated sea glass piece like this one is even more incredible.

This beautiful example of colored American glassware was found by my friend, Iris, about twenty years ago, also on a remote Pacific Northwest beach.

If you have uniquely colored sea glass that is pink, yellow, dark lavender, teal green, or creamy blue and if your piece shows a decorative pattern, it is highly likely that it originates from early, decorative, or depression era tableware.

This bodacious, diamond shaped pink piece is very
obviously a top handle of some kind. Usually connected
to a bottle stopper or dish lid, this color is commonly
seen in early 1900s tableware. Pink is considered highly
rare wherever it may be found on the planet. More about sea
glass color rarity is found on page 133.

This incredibly vibrant, lime green handle looks exactly like the turn of the century cut-glass doorknobs. However, this one is made of pressed glass and it does not possess the hollow for the metal mechanisms needed to turn the latch. Pressing doorknobs in molds was first started in 1826. I do not believe this piece is that old. When I first found it, I shared with fellow sea glass lovers that it may be the top piece to King Triton's under-the-sea scepter. I do believe that it could very well be an ornate, glass top knob to an antique walking stick.

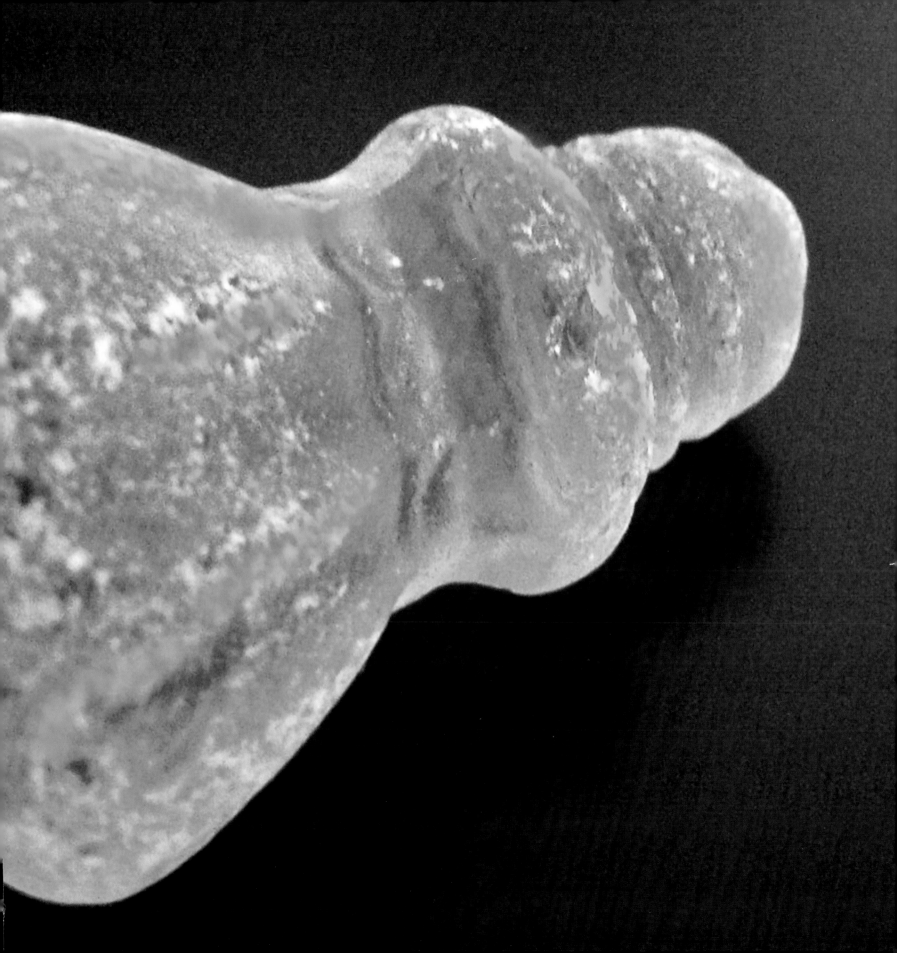

The scalloping design on the edges and the beautiful
teal color of this perfume bottle would have had
me guessing it was a vintage pickle jar. But the size
of it (about three inches) and the intact stopper at the
top indicate that it is likely an early American perfume
container.

Here our traveling display shows some beautiful pieces of antique tableware from Rhode Island, Massachusetes, and New York. We have discovered that the older pieces from Atlantic shores tend to retain some of their identifying marks unlike similar pieces from the West Coast where patterns and definitions are tumbled away.

bonfire glass

Incinerating garbage piles has been common practice ever since humans have desired to remove refuse from any given location. Since many garbage dumpsites used to be located along our beaches, it was not difficult to find incinerated glass and bottle pieces left on the beach after such a burn pile had been extinguished.

Bonfire glass is a result of this incineration practice. Bonfire sea glass is glass that has been found after it's been melted and incinerated in a beach bonfire. It is typically found along beaches that were located in proximity to settlements and towns.

This beautifully conditioned and ocean-sculpted bonfire glass teardrop-shaped piece of sea glass clearly shows two separate bottle colors; aqua blue and amber brown.

Usually bonfire glass is unattractive and gnarled with curled edges that are bent and twisted over from succumbing to the melting heat. It often melts into and onto other pieces of garbage, metal, and glass that also are in the burn heap. It can attach and fuse to all sorts of inorganic material, organic material, and even rocks and sand.

Normally time at sea and in the elements will biodegrade away the garbage that did not burn in the fire and the resulting remains are the glass chunks that originated from discarded bottles or tableware.

These glass chunks are left on the beach to tumble in the waves, over and over for years.

At right is one of the most amazing bonfire glass pieces I have ever seen. It could be misidentified as a piece of fused art glass but it is not. It is a near perfect bonfire fusion of two separately colored bottles; cobalt blue and frosty white. The beauty of the piece is that it is fused together right at its middle point with almost two equally sized and shaped sides.

Additionally, this piece has had all of its curled edges and pitting "sanded" away by the time it has spent tumbling along the shore; it's a sure one-of-a-kind piece.

sea glass beads

Not to be confused with tiny pieces of sea glass that we find then drill a hole through, a true sea glass bead is an historic, once-in-a-lifetime find even for serious hunters.

A *sea glass bead* is an actual glass bead that was created to be threaded onto something; therefore, it has a man-made hole through it.

These uber rare discoveries were actually glass beads that decades earlier had adorned a lampshade, hung from a woman's jewelry, or were used for other decorative purposes.

Though it's rare to stumble across them, some beads found on the shoreline used to be a type of currency among coastal tribes and journeying seafarers for supplies trade. We believe the darker, oval-shaped bead in the background is in fact a trade bead. We always encourage the collector to return any Native American artifacts to the regional tribe's museum or representatives.

Usually older than bottle glass, which began washing ashore after mass bottle industrialization in the early 1900s, beads are likely handmade, one of a kind, and sometimes originated from far-off lands. In fact, glass beads are said to be the first things ever created out of glass.

Therefore, by deduction, it seems true that the collector who finds sea glass beads is usually hunting in fairly historic locales.

By the late 1950s, plastic beads were replacing glass beads in US households and in many crafters' palettes. So it is quite likely that if you find a true sea glass bead, it may very well be from the early 1900s or before.

This rare bead is not only made in the most desirable color (true orange), but it is also well frosted and faceted; three factors that place it high in value and beauty. It was found in the Pacific years and years ago. It is well frosted by decades of tumbling along a rocky, Northwest shore. We believe this is one of West Coast Sea Glass's rarest beads. While trying to date it and uncover its origin, we've come to find that it most resembles an early 1900s antique, a Czechoslovakian "nailhead" bead. These are flat on the back with faceting on the front and sides. These were used one hundred years ago and in the days of the flapper to bedazzle clothing before sequins were created.

This exceptional red flower sea glass bead (or button) caught my eye while I was judging the most recent Sea Glass Festival Shard of the Year Contest. Rarely do we see such a bright, cherry red colorant in sea glass. This piece shows a near-flawless surface with uniform conditioning. It was found by collector Cindy from New York.

This red teardrop bead was a big surprise to find. We have a few, small *round* sea glass beads in red. We've traced them back to fishing rods and tackle. This bead, however, is not round like one would see on a fishing pole. It is a true decorative, teardrop bead; our only one in red.

a child's discovery

Every piece has a story. When my twins were toddling about, I took them to the beach often. We adventured to shores throughout the Northwest United States on a regular (sometimes daily) basis. Little Emma would sift through the sand with me while we built sand castles. She'd use seashells, seaweed, and driftwood as adornments on her palaces. Occasionally she would use a piece of sea glass from her pocket too.

One afternoon when the twins were about two-and-a-half years old, we had come back in from our local beach for naptime and I had tucked them into bed and was giving kisses when Emma held up her stuffed animal toy (a baby chick with a zipper pocket) to me and said, "Mama, look in my chicken's pocket." I reached in and pulled out the most vibrant, brightest yellow sea glass piece I had ever seen! And it was a truly perfect, sea glass bead. I couldn't believe it!

My beach is not known for very much sea glass at all and it surely is not known for rare pieces. I asked her where she got the piece and she answered, "From the beach." I never learned exactly what beach she found it on since she toted that stuffed toy around for most of a year.

She didn't care much about the bead. I suppose she viewed it as about as important as one of the seashells or driftwood pieces she left back on the shore. She gave the bead to me that day and since then it has traveled around the country as a featured piece in our museum display.

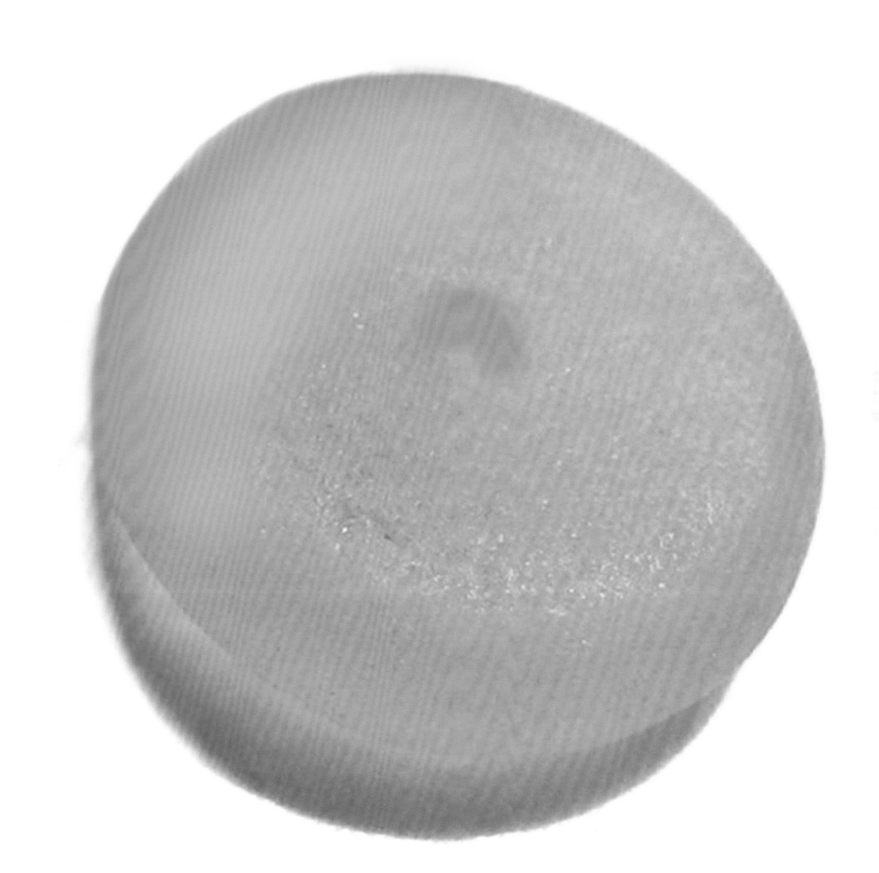

pieces with swirls of multiple colors

bonfire glass

As you've already learned, bonfire sea glass is glass that's been incinerated and the heat melts multiple vessels of varying colors together.

flash glass

Another type of multi-hued glass is *flash glass*. This is not bonfire glass. Flash glass is manufactured to have multiple colors usually of a base layer with a thin contrasting layer of another color flashed or fused to the surface.

sea glass marbles

Other multi-colored sea glass simply originates from refuse glass originally created specifically to have multiple colors in the piece by a manufacturer or art glass blower. This is not flash glass. It is art glass that had various colors in it in the first place. Art glass sea glass is discussed on page 210.

In addition to these three common multi-colored pieces, another type is created when multiple separate colors are blended together while the glass itself is in the liquid process. Here, the residual molten batches were mixed together once various colored work projects were completed in the factory or studio. They coalesced together in clumps, pieces, or globs, then were "pitched" into the sea. The surf then tumbled the colorful shards smooth.

A trio of sweet egg shaped sea glass pieces with gradients of beautiful multi-colored layers.

Swirls of mixed colors blend beautifully in these sea glass pieces beachcombed from the shores of England's North Sea. These particular pieces have been in the spotlight in the sea glass world for much of the past two decades. Some collectors have even termed this glass "end of the day" glass because sometimes the colors have been mixed together at the end of the project which, at times, was completed at the close of the day.

sea pottery, ceramics, and porcelain

If you are an avid sea glass collector, it is likely that you also have a few shards of pottery, ceramic, or porcelain in your collection. Sea pottery is synonymous with sea glass. In fact, most collectors keep their ocean-tumbled pottery collection in with their sea glass collection.

Sea pottery basically originates from pottery, crockery, and dishes that were discarded into the sea. Items break into smaller and smaller pieces and are tumbled smooth by the wind, waves, and sand along the shore.

Sea glass pottery is more porous than sea glass, making it quite a bit more delicate than a heavy nugget of hard glass.

The photo at the right includes several
recognizable Fiesta Ware pottery shards in pastel
colors, which were popular from about the 1930s
on. Fiesta Ware and other similar pottery lines came
in many soft shades which are recognizable here—
including the highly sought after pieces with stripes.

Patterned transferware (decorated using a transfer technique) and blue and white pottery, including porcelain dishes, were very widespread and were among the most highly collected kinds of dishes and tableware in the United States and Europe. The original dinnerware patterns were first created in England in the 1820s. Styles became very popular in the US in the early 1900s.

dolls and toys

Doll and toy parts were also made of ceramics beginning in Germany in the 1840s. Bisque ceramic, with a more matte finish were created and used especially for doll parts. It was very widely used throughout the toy market after 1860.

The porcelain statuary child's head at left was a rather unnerving entry submitted into the "figural" category at a recent sea glass contest I judged.

Australian diving records from 1875–1891 archive several items uncovered from the wreck of the *Figi*; a three-masted iron barque from out of Ireland. Some of the items that divers have spotted at the wreckage site have been recorded; gin bottles, ceramic toys, and broken doll parts. Porcelain and pottery can weigh much less than a piece of glass its same size, and many toy and doll parts are hollow inside, which give them a buoyancy advantage.

The sea glass collector who finds a porcelain or ceramic toy or doll part has come across a highly rare novelty.

An indication of just how ultra-rare they are; I have only one piece in my collection, pictured at left. I have identified it as a tiny doll arm, measuring about 1.25 inches.

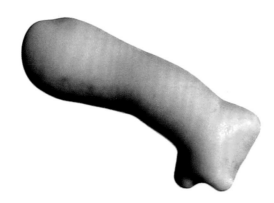

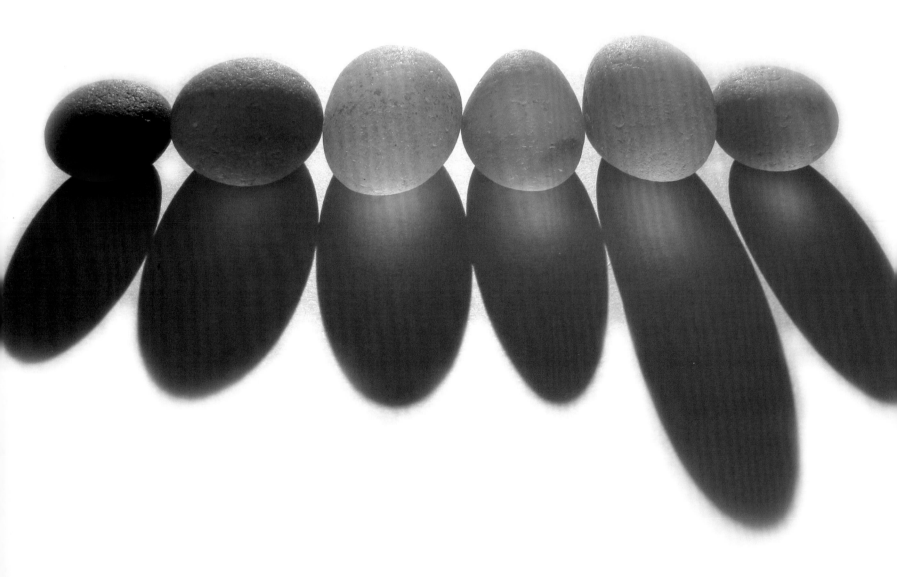

PART THREE

COLLECTING

the international sea glass community

For several years in the early 2000s, I was the main moderator for an online discussion conglomeration of sea glass aficionados from all over the planet. We functioned under a simple name: Sea Glass and Beach Glass Collectors and Enthusiasts.

The magic of that online group was that knowledge was freely shared across a wide expanse of collecting experience. And we truly became a community. Some friends checked in weekly, while others became daily visitors to the discussion site. Though we had several hundred participants, the locations represented that I especially remember were (in no particular order):

- Nova Scotia, Canada
- Prince Edward Island, Canada
- Lake Superior
- Lake Erie
- Sussex, England
- Netanya, Israel
- San Francisco Bay Area, California
- Florida's east coast
- Florida's west coast
- St. Thomas Virgin Islands
- Puerto Rico's Atlantic and Caribbean shores
- Kodiak Island, Alaska
- Rhode Island coast
- Massachusetts Bay
- Hawaii, Big Island
- Central Oregon coast
- Long Beach Peninsula, Washington
- Honshu, Japan

Collectors and experts from all over the world converge on sea glass events. Here, some judges and volunteers deliberate over which pieces are the most prized, rarest, and most beautiful.

Since I was as close as I could be to the pulse of the sea glass world, I'm positive that our small community was the first of its kind. I became determined to help lead the way to keep these amazing comrades connected from then on.

We would discuss our finds and often we'd share photos and help one another identify the origin of pieces. We'd collaborate and agree on the names of colors and if the pieces were "frosted" enough or if we felt it should have been "thrown back for more conditioning." We'd sharpen one another's knowledge of the subject and share as much of ourselves as we were comfortable with. We had respect and appreciation for the next person's journey, which often was considerably like our own.

For some of the collectors, sharing that journey of discovery was a gift to the receiver. But for others, sharing what they knew and had experienced could almost be too personal, like sharing a piece of oneself. When you are a sea glass collector, you understand this and hold it high. You understand how sacred for some the voyage of a human being plus a beach—plus a quest—can be.

The group from the online community shared and stayed active for several years and the endeavor was fun and much more informative than I had imagined it could be.

The group grew and the volumes of information grew with it. The task of keeping every question answered and every story told became like the ocean itself—a deep and wide project. It was right about this time that several of us decided that this info and growth was proving to be an exciting time. We were being pointed into restructuring and now was the time to build an actual North American group that we called the Sea Glass Association.

Onlookers wait to hear whose entries won the various categories in a shard contest.

The North American Sea Glass Association

Our first tasks were to write goals and bring the community together with information and festivals with the purpose of providing displays, lectures, and a chance to show off one's collection.

We put together our first official event in 2006 and held it in Northern California. Thousands of people came from all across the United States as well as from Canada, Ireland, and England. The most frequent comment I heard from the attendees at that event was "I had no idea people were into sea glass like I was!"

I was elected president of the organization and served for another six years. And now, here I am, twelve years since my first experience with the "sea glass community." I am still walking in the early mornings on my home beach along the Puget Sound and I still travel to other beaches as often as my schedule allows. My goal now is to catch the sunrise, find some special pieces, and catalog a more personal, simpler odyssey.

no borders

The sea glass community has developed and branched out across the globe. And now even broader reaching organizations exist that stretch to the ends of the earth like the International Sea Glass Association. The goals are the same as ours were—to educate and connect the collectors of the world.

a tight-knit community

Though we are now spread all over the world, we are still very connected. There have been a few times when I've wondered to myself, "Wait. Is this sea glass world really about sea glass or is it really about people?"

One of the best examples I can think of that points to the beauty of this group is a very personal one. It was early September 2012. The national convention was just a month away, and I had purchased my plane ticket to the East Coast and hired staff to travel with me to host the West Coast Sea Glass booth. This event is usually our biggest undertaking of the year.

Unfortunately, that same month, I had become ill and doctors had found and removed a large cancerous tumor. The colon cancer had spread, so I started what would be a long seven months of debilitating chemotherapy. There was no way I could attend the big event.

My coworkers, Lindsay and Janet, had to tackle the trip without me. A few phone calls were made between them and the event managers, some which I was to overhear from my bedside.

Though I wished I could be there with them, I was thrilled to hear the trip and the event were a success. Soon after their return, Lindsay was putting away paperwork and supplies in the office, and I was sitting there watching and probably doing my occasional office-chair directing but not lifting a finger. She was gathering two piles into two envelopes—one of them bulging and spilling over. She handed them both to me.

As I inspected the contents of the first envelope, I found it was filled with cash and checks made out to me personally. This heap did not represent our business profits from the festival; it consisted of monetary gifts from other exhibitors and vendors, all for me to put toward my treatment and overwhelming medical bills!

The other, more mountainous stack in the bulging envelope turned out to be dozens and dozens of get well cards, notes, and sentiments that went along with the gifts. My sea glass friends, some of them my direct business competitors, were praying for me, thinking of me, encouraging me, and wishing me well!

They had organized a massive fundraiser in my absence; each exhibitor had posted a small info card at their booth with my photo and my story on it. Each had earmarked the sale of special items that would to go toward my support. And those checks and notes were what Lindsay brought back and lovingly offered to me from the group.

Hundreds of sea glass pieces from all over the world line the tables during a big sea glass event. This particular contest has nine categories that can be won.

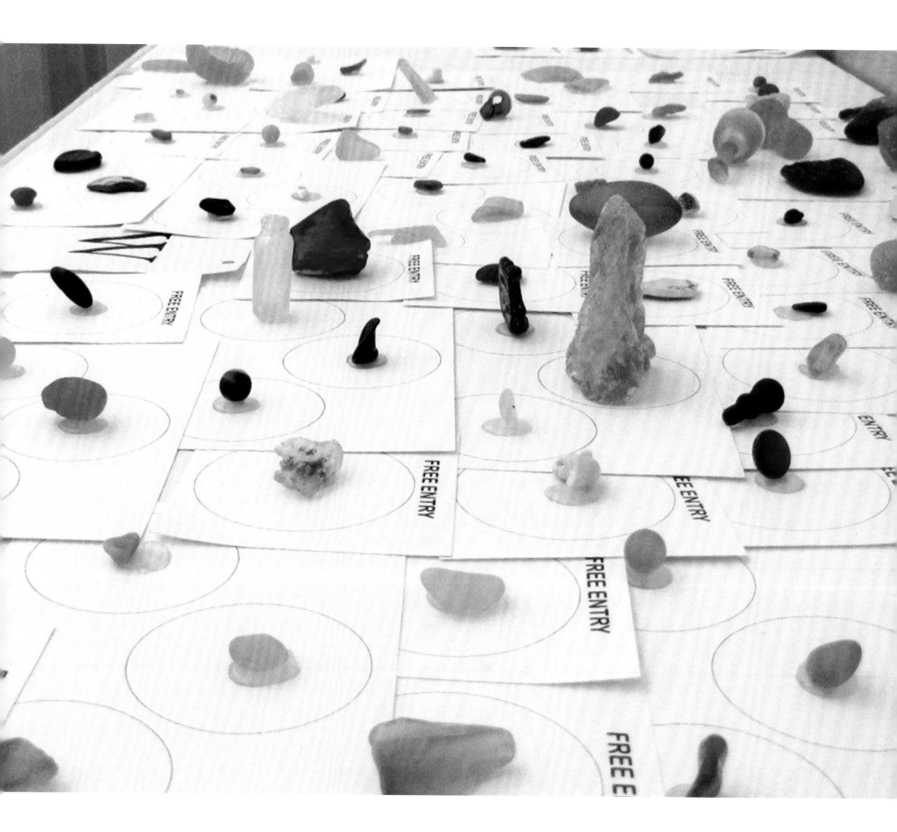

Little did I know that the sea glass friend who originally proposed the idea is one of my most far-away collecting friends in Great Britain. She spoke to the board of directors, whose early phone calls put the wheels in motion as they schemed to implement a fundraising event on my behalf.

And the gift keeps on giving. A year later, I was having dinner in Pennsylvania with two of my sea glass friends from Maine, Lisa Hall and Julie Havener. They are first class women! If you know sea glass art, you know their work. It is some of the best in the world. Lisa would be mortified if she knew I mentioned this, but I will mention it anyway (and will text her later that I did—of course!). She had been extremely generous with her cancer fundraiser gift. Though all of the donations brought tears to my eyes, it may have been hers that brought the waterfall down.

While dining that evening in Pennsylvania, we somehow got to talking about the previous year's event, the one with the table-top fundraiser. She and Julie (who also traveled a great distance to the annual convention) had commented that though they made a strong income, it still wasn't their most profitable of events. Likely forgetting that she had made a large donation to my cancer fund during that very same event, she mentioned how much her company did make. I quietly sat there in the loud restaurant and did the math in my head, realizing that she donated over 80 percent of her profits to her friend on the other side of the continent!

I can think of no better story to explain how beautiful most sea glass people can be.

A bracelet made of beautiful blue sea glass. Lisa Hall's jewelry has included pieces from the West Coast Sea Glass collection for nearly a decade.

the collection chart

More than a decade ago, I had grandiose dreams of keeping track of how many pieces I would find over my next twenty years of collecting. I envisioned making a detailed chart with graphs and particulars delineating all the factors that might affect the amount of glass and type of glass my collecting friends and I found. The chart would tell the world of the nuances of sea glass collecting and how several factors can greatly affect one's "luck" during the hunt. The idea was and still is a good one. Keeping the chart current and accurate over a twenty-year period, however, was the challenge.

There are many key factors that influence one's sea glass hunt, but the most influential ones are as follows:

- Beach location where you are hiking
- Where the tide line is on that particular day
- How far one hikes
- What the weather was like prior to the hunt
- Quantity collected
- Rarity of pieces collected

The chart sat in my journal in the drawer next to my bed for years. Though it was close at hand, it received intermittent attention. Sometimes the sea glass collecting logs were a day or two apart and some entries fell weeks apart.

I kept the log loosely going for about four years. Keeping the log was purely for informational purposes and was a sort of recreation for me. I enjoyed gathering interesting facts as I worked, and I never would expect my cohorts to be as into it as I was. It was a pleasure to look back on and remember each beach trip and journey and experience. Though I had made a point only to record my personal finds, the log also helped me recall the more personal moments with my friends and family.

I called the chart the "written chart" because, in all honesty, I still walk with a chart such as this in my head—my unwritten chart, if you will. It resides in that junction of my brain that is always logging and always calculating and always enjoying my sea glass experiences.

A sampling of the written chart read like this:

Date	Tide	Location	Wind	Dist.	White	Brown	Green	Cobalt	Aqua	Rare
6/18	5.3	Central Oregon	8 mph	2 miles	16	9	6	1	1	0
7/11	1.6	Baja Mexico	3 mph	1.5 miles	2	7	4	0	0	0
11/23	0.6	Kauai, HI	5 mph	2 miles	7	18	12	0	4	1
12/2	3.22	Strait of Georgia, Canada	9 mph	1 mile	3	8	4	0	2	0
12/25	2.6	Dungeness, WA	1 mph	2 miles	3	12	0	0	0	1

If you, too, are a serious sea glass collector, I will assume you can understand how I think. We are the type of people who cannot help but remember a special afternoon, which leads us to remember a particularly distinct hike, which reminds us of that exceptional piece that we found that we will probably not forget. Most serious collectors can remember where they were and when it was when they found that memorable piece.

In fact, as I review the chart, I can remember the specific piece I found in Kauai on November 23. My parents would have been celebrating their twenty-third wedding anniversary that weekend. My father had recently passed away, so my mother decided to invite all her children to Hawaii for a once-in-a-lifetime trip.

make your own sea glass chart

Creating your own way to log your collection habits is easy. Tailor the graph to the environmental factors that will affect you in your particular area. From my studies, the most influential determiners are:

- location
- high tide mark
- distance walked

But you might want to include other factors like what time of year it is, how many collectors are in your group (sometimes children in tow can add to your collection or you might end up spending the afternoon playing in the water instead). If you care about color and condition of glass, leave room on your chart to also include those factors. Gear it toward you and your habits and you'll enjoy the graphing process even more.

My little brother and I had been walking on Poipu Beach and we came upon a very large, thick teal and aqua blue piece of old electric insulator glass toward the end of our beach walk.

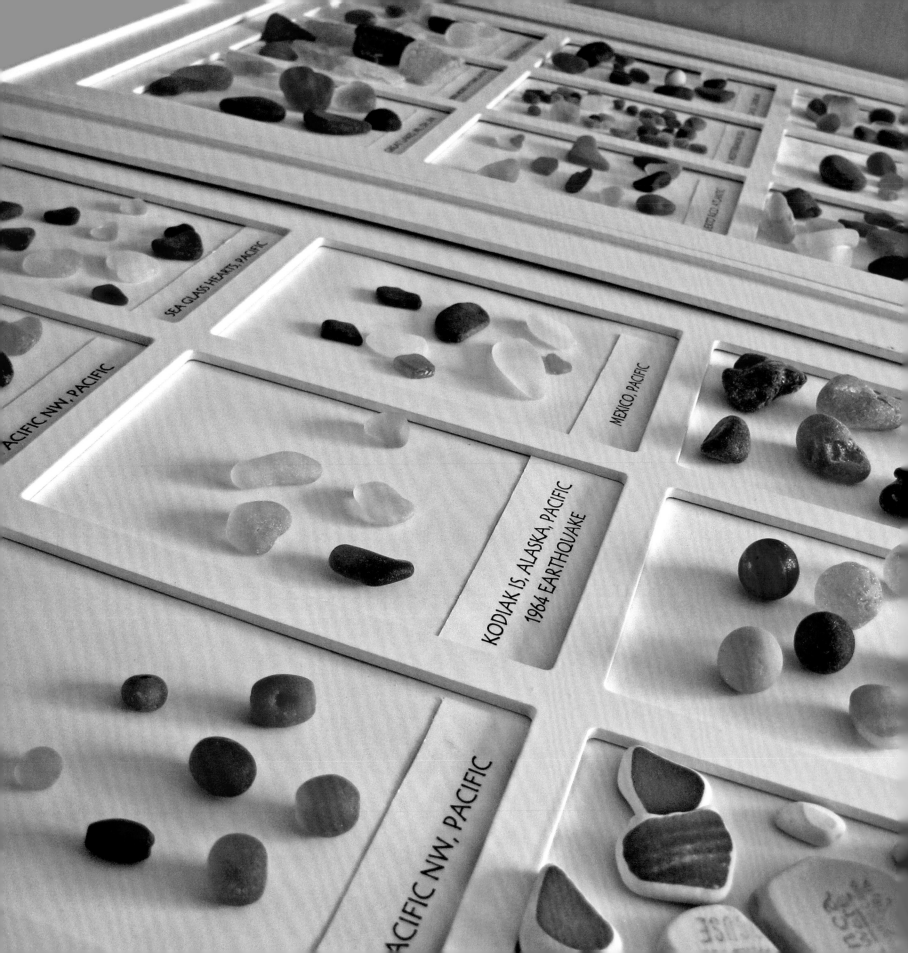

when and where to go

At my house we live by several charts—the calendar helps us keep our appointments and activities. The barometer assists in knowing what the weather might be doing in the next few hours. And the tide charts tell us when the best times to head to the beaches are.

If you're going to go out collecting with any resolve, feel free to use these helpful tips:

Check the tide chart to find out when the next couple high tides are. Your very best day to head out is on a King Tide day. A *King Tide* is the same thing as a *Perigee Tide* and there are only a few of them annually. But there are many days a year that have what I call *dramatic tides*. Simply explained, they are any day that the tide moves high up on the beach, then recedes far down past the usual median wave line.

Make sure you head out with the first daylight after the high tide. On the perfect day, you will walk as the tide is going out. This will ensure that you are present directly after any sea glass has been "dropped" onto the beach by the last high wave.

Head out after high winds and storms have come through. Storms tend to churn up sand and beach that has been dormant previously. Hopefully, the storms will also churn shoreward any sea glass that has been covered and hidden previously or buried offshore beneath the sand's surface.

If it helps, keep log of your finds paired with tide behavior. A collection chart like I described on pages 104 and 105 should work fine. This will help you realize the correlation between your results and how the tides might provide more glass.

Find an area where historic settlement has occurred. This will increase your chances of finding older pieces that ended up there when coastal dumping was allowed.

Locations that are farther north on the planet are known for more dramatic tides. Find a beach that has good tide fluctuations in height. This moves glass nicely and smoothes the edges over time.

Rocky or pebbly shores are best as they have been proven to "catch" sea glass better than wide, vast, open beaches. Pebbly shores acquire pebbles and sea glass behaves in much the same way. If your beach has a rock wash of pebbles and small stones, these patches will be where sea glass conglomerates also.

When you walk, stay in the strand lines. A *strand line* is another name for the high tide mark sometimes used to describe where a flood line has reached. These are noticeable as you look ahead and down the beach. They are usually filled with myriad deposit material like sea shells, seaweed, small pebbles, and small pieces of garbage debris that hopefully include sea glass. You will notice that some wider beaches may show several strand lines on any given day. The heavier, larger pieces of sea glass (if there are any) will be up higher on the beach within the highest strandline. The smaller, lighter weight pieces will have rested slightly lower on the beach, most likely along a less prominent strandline.

Find beaches that aren't teeming with tourists and hunters who amass large collections.

If you're able, find locations that have a history of marine traffic and that land or used to land along major marine stops. These sites tend to avail more pieces and remnants that have a more sophisticated or unique past.

tides and debris

It was summer in the early 1970s. I was in grade school. My family of nine (seven kids) was on our way to a vacation to British Columbia, Canada. We took a ferry boat across the Juan De Fuca Strait from the United States. I don't remember the reason, but there was an inordinate amount of partying young people on the ferry boat during the trip. I recall my parents commenting on how many empty liquor bottles were left around the deck and in the galley area.

As a result, a few of my siblings and I got into a discussion about picking up the bottles and using them to send into the sea with messages in them; messages in a bottle. It was then that we decided we'd run our own experiment. Each of us would recycle and reuse one dumped bottle. We'd each write a note; being careful to make sure we left clear instructions to the lucky finder to correspond with us when they found the message. We each stuck our notes inside our respective bottles, corked it, and then tossed it off the ferry boat into the sea. This particular stretch is about twenty miles wide.

Our family vacation started, came, and went. We traveled home and every single one of us had forgotten about our messages in the bottles. Weeks passed. None of us ever heard anything of our bottle journeys until . . .

Twenty-six days after I tossed my bottle off the starboard side of the ferry boat, I received a letter from a little girl named Marissa. Her uncle worked a farm on Lopez Island; one of the San Juans. It was there that they came across my bottle on the beach. The shore they found it on was about twenty-two miles northeast of where I dropped the bottle. Marissa and I stayed pen pals for several years.

morning hunts

The currents, winds, and tides took that flotsam twenty-two miles in twenty-six or fewer days. Marissa's uncle said he found it in the morning after high tide.

As we've discussed, the best time to head to the beach for sea glass collecting or beachcombing is after high tide has receded. Mornings seem best if you want to get the fresh deposits from the night before.

The moon's pull on the earth creates the daily advances and retreats of the water's tide. This is what changes the shore's tideline throughout any given day.

king tides

The best time of the year for sea glass hunting is after a King Tide. *King tides* are the highest seasonal tides that occur along the coasts of the planet. They represent the largest tidal range seen over the course of a year. These impact our shores by rising waters. Our shores are constantly altered by human and natural processes and projections indicate that sea level rise will exacerbate these changes.

Currently there are groups working hard to record the changes to our coasts and shorelines by photographing the king tides, which only happen six times a year. This project is giving us a glimpse of what our daily tides may look like in the future as a result of sea level rise.

The king tides occur three times (once a month) during the winter months and three times (once a month) during the summer. These tides can greatly affect how high or how low on the beach the waves can reach. Some places on the planet on certain days can see as much as a fifty-foot tide gradient. The Bay of Fundy, on the coast of Nova Scotia, Canada, has been said to have one of the most extreme tides in the world.

Our most dramatic tides can fluctuate over nine feet where I live along the Olympic Peninsula in northwest Washington State. A piece of sea glass can cover many more miles along a shoreline like this than along a shore that gets little to no tidal movement. A shore like this, with what I call *dramatic tidal movement,* can frost and tumble a shard from a sharp-edged piece to a smooth, premium, well-rounded bauble.

You can see from this photo where my children are digging for clams that the tide is very far out. The houses in the background are located where the high tide line usually is. This is one of the reasons why we are out digging clams; it was one of the King Tide days.

lesson learned

I grew up along Oregon's Pacific Ocean shores and spent my childhood with an experienced understanding that the water level appears at a different spot on the beach every morning. Even if I were to awake at the same time daily and look out the beach cabin window, the highest wave point would be at a different spot than it was at that same hour the day earlier. This is because every day the time that the tide comes in changes due to that day's pull of the moon's gravitational force.

I lived in Oregon for the first twenty-seven years of my life, then I moved to inland Washington State and lived a land-locked life for about eight years. It was when I was a new Washingtonian that I was invited one summer morning to a day and dinner at "the lake" cabin with friends. We had a barbeque, threw the Frisbee around, and swam a bit that afternoon. Later in the day, a nice calm came over the lake and I noticed an empty canoe near the shore. There wasn't much beach, just a narrow trail, really, between rushes. I asked my friends if it was their canoe and if I could take it for an evening row. They said sure.

I gathered a life vest and oar and, as I pushed off shore, I hollered back toward the cabin and asked, "What time is low tide tonight and how far will I need to pull the canoe back up on shore?"

Everyone at the party standing on the cabin deck stopped and looked my way. There was a pause, some blank stares, and then chuckles.

"Oh, that's right. You've really only been around the oceans!" my friend Phil laughed.

Up until that point in my life, I hadn't really been around lakes very much. I had not put much thought into the fact that tides don't fluctuate on smaller bodies of water like inland lakes. It was a good lesson and, once I moved back to live along the Pacific Coast again, I was reminded how wonderful it is that the tides shape the shore and give us a new beach almost every day. The fluctuating tides along any ocean's shore will bring in everything from sea glass to other garbage, driftwood, derelict boats, and more.

effects of natural disasters

The Oregon State Parks system has a very intricate beach debris cataloging and disposal system. As we've discussed, much of this debris originates from a great gyre in the Pacific known as the Great Pacific Garbage Patch. Activity in the Oregon State Parks program has increased greatly as a result of the large 9.0 Honshu, Japan, earthquake and subsequent tsunami in 2011.

During the keynote lecture at the 2013 North American Sea Glass Festival in Virginia Beach, I showed photos and discussed in great detail the kinds of debris washing up as a result of the tsunami, currents, and tides since the quake. The California, Oregon, Washington, and Vancouver, Canada, coasts are showing signs of Japanese debris.

Sea glass hunters can learn much about how this natural phenomenon has affected what is washing up along the coasts. When we travel and show our collection at northwest shows, one of the most common questions we get asked is: "Is that big tsunami from Japan washing up a bunch of Japanese sea glass?"

I try to answer kindly, "No, because glass does not float." But I also add, "Yes, if it is a buoyant glass piece like a glass float or closed bottle." But once that glass reaches the rocky, West Coast shores, it likely has dashed upon the rugged shoreline and the pieces will sink to the sand.

We are discovering that the larger, more buoyant items are what have washed up first from the big tsunami; these items tend to travel faster upon the surface of the sea and along the faster current(s). I lectured about how an entire dock from Japan had washed up on the West Coast. It measured 65 feet long, 20 feet wide, and 7.5 feet tall. I had also read about an entire motorcycle washing up as well as large buoys and entire boats.

My friend, Cliff, who walks his dog daily around the Olympic Peninsula, came upon this large Asian buoy float in La Push, Washington. The La Push beach is in the top most northwest corner of the contiguous United States. This is a perfect example of a larger, buoyant item that is not glass and that has crossed the sea on a rather quick trip.

In the fall of 2013 the National Oceanic and Atmospheric Administration's Marine Debris Program was seeking 150 soda-bottle GPS drift transponders. These are research bottles that were dropped into the water off the coast of Japan to study drift patterns since the March 2011 earthquake and tsunami. The bottles have been floating among the debris for more than two years, and some are expected to land on beaches in Washington State, Oregon, California, Alaska, and British Columbia.

Beachcombers are asked to watch for these transponders, which were put into the water three months after the tsunami, and follow directions on the bottles to contact researchers. So if you're a beach hiker, take note. The findings should help us to understand currents and tides even better.

caring for the shore

As a serious beachcomber, it is important to know that beach regulations vary from country to country and state to state. For example, I grew up in Oregon as a child. That was during a time when the vast, wide beaches were considered public property. Once we had access to the beach, we were allowed to roam free for miles and miles. Then I moved to Washington State where the shore access rules were much more rigid. Most waterfront homeowners own the shore in front of their property up to a median tideline. I had a whole new set of regulations to be aware of.

It is considerate to be mindful of local beach guidelines and laws. Remember many communities have private property laws and it is wise to show respect to the locals who live and work along the shoreline and who consider the beach their home.

It is also wise to educate yourself about the environment and how to best care for the earth. Beach lovers are a breed of humans who usually have their ear to the proverbial sand. They are usually fairly earth-minded. So being aware of shoreline restoration projects, wildlife habitats, property lines, and tide lines is always a conscientious move.

When my friends, family, and I beachcomb, we always make a point of being sure to bring a garbage receptacle with us on our treks. Our philosophy is to always leave the beach cleaner than when we got there.

Lastly, it is decent to be sure to respect the discovery and solitude of the adventure. Some beachcombers love the tranquility of nature and especially the meter of the rhythmic waves. Keep in mind that much like how a fisherman feels about his favorite fishin' spot, a favorite sea glass hunting beach can be like holy ground to some.

further tips on collecting

Where can I find sea glass? This is a question I am asked often. There are several reasons why the hunt is so important to the seeker. Some find their Zen moments in the solitude and beauty of the hunt. Some collect to add color to their life. The history, mystery, and discovery is also a strong force that draws the collector. Whatever your reason, the sea glass collecting window is closing as pieces are becoming more elusive.

One good tip to keep in mind once you do find a beach is to collect where the beach pebbles of certain size conglomerate and where the finer materials of seaweed and smaller splinters of seashell line up along the sand.

The seeker can walk a mile in one direction along a shoreline, stop, turn around, and walk the entire distance back, and on the return trip he or she will notice many pieces of sea glass that were missed.

This is because the sun can play with your eyes as you cast them to the beach. Some pieces of sea glass will hide in the shadows (especially on a pebbly beach) so they cannot be seen when approached from one direction but are much easier to see when approached by the opposite direction. In fact, some of the best weather to collect in is when it is overcast. Sunglasses can be one of the best pieces of gear for the beachcomber on sunny days.

Be mindful that for many, sea glass collecting is a challenge that's defined over sometimes decades of combined seasons and years of hunting. For some collectors, the hunt is about covering countless beach miles by foot and kayak. And the joy of refining the searching experience itself is part of the allure, the mystery, and the journey for each individual seeker.

Be sure to enjoy the providence that seems to ebb and flow as you seek.

Part Four

IDENTIFICATION

different beaches and oceans condition sea glass differently

Most sea glass collectors will tell you that the frosty patina on the surface of sea glass is what defines it as, in fact, sea glass. But can that frosty surface look and feel different depending on the environment or beach the glass was found along?

The answer is yes. How noticeable of a difference, though, can be very tricky to distinguish.

This thumbnail-sized, emerald-green oval at left was photographed on a Malibu, California, beach in the exact pebbly shore location where I found it. This piece of sea glass took on the overall shape and smoothness of the rocks and pebbles it was found amongst.

I posted the photo on the right in an online photo album and it garnered a lot of discussion. One gentleman and I went back and forth regarding the "genuineness" of my pieces. He had been a collector for a couple years, collecting daily, and he had a nice assortment with hundreds of pieces. His concern was that he had never found pieces so uniform in condition, size, and shape. He was sure they were man-made and not the real deal because none of them had rough edges or seemed "flat" like bottle glass.

I asked him about all the different locations he had collected and he shared that he'd only beach-combed along his hometown beach at the edge of Lake Superior. I was able to gently explain to him that my pieces originated from glass that was tumbling along a much rockier shore than Lake Superior's and that they are from a site where colorful, art-glass dumping was done more than two hundred years ago. We surmised together that the terrain where my pieces had tumbled also had substantially larger waves and tidal fluctuations than his local shore.

Where I live, my home beach is inside a bay that is nicely protected by one of the world's longest sand spits. We rarely have rounded, smooth, ball-shaped sea glass on the three-mile stretch of shore that I walk most often. But there is some history here; the pier pilings at the end of my road tell of a time in 1890 when a small port was built to serve the growing trade in the newly settled area. It did not take long for me to conclude that my local sea glass was not very well tumbled primarily because my beach lacked the fiercer wave action that the more exposed shores on the outside of the sand spit received.

It is true that beach terrain, severity of wave action, and even tide depths are key factors in conditioning a piece of sea glass.

The Caribbean Ocean flats on the left show some of the more common colors of sea glass; olive green, amber brown, and forest green bottle glass. The interesting feature to me is the silkier patina on the surface of the pieces. The pitting in the glass surface is less deep and lacks the more mottled behavior that is commonly seen in North Sea or Mediterranean Sea glass bits that I have observed.

From my experience, glass from the Great Lakes and the Caribbean have a more silky surface than sea glass from other world locations.

a sea glass conversation

One afternoon we were sorting similar hued cobalt blues to use in some earring designs. We had pulled together some pieces that represented several lifetimes of collecting all up and down thousands of miles along the Pacific.

Here's a direct conversation I had with three of our sea glass "students" about how smooth the pieces looked:

Jeanne: That's a lot of smoooooth blue. Please tell me where I can find that much.

Judy: Wow, M. B.! So much blue! I'm drooling!

Mary Beth: Jeanne, those blues cannot be found in any one particular place any longer. They are a conglomeration of years of hunting, from about sixty years (combined) of four people collecting all over from Mexico on up to Alaska—all Pacific Ocean in this batch.

Julie: Does the location of the glass have any affect on how it turns out in the end? Do some oceans make it smoother or cause a different level of frostiness? Or is it all about the same and just depends on how long the glass has been tumbling?

Mary Beth: Oooh, these questions are super! Yes, the ocean does matter and the location matters. Rockier shores seem to avail frostier, better tumbled glass. Tidal and wave movement can make a big difference too. Some places north on the planet have forty-plus feet of tidal movement (up and down) and this can really "rake" a piece along over a couple decades. I've found that water can "seep" into glass as evidenced by some of the glass floats we've seen, and we also have a few sea glass pieces with water in them. My more scientific friends have concluded that sea water, salt content, and acidity can make a difference too.

A customer sent me some glass last week from another part of the world and I was able to guess correctly where she had beachcombed it in Spain just by looking at several factors; smoothness, size of the pieces, color, etc.

Jeanne: These are my favorite color! I feel like I'm looking at gem stones instead of glass, they are just so bright and beautiful!

Color Rarity Scale

In 2001, I developed my first sea glass color scale based on rarity. I have added more pieces to the chart but, for the most part, it has remained a constant for collectors all over the world who want to reference their sea glass scarcity. It represents a fairly generalized worldwide scale.

Sea glass rarity is often about the uniqueness of a piece's origin, but rarity is mostly determined by color. We've studied tens of thousands of pieces, collaborated with sea glass experts, and spent some serious years hunting the globe on our own to develop the reference. It reads from left to right, top to bottom. Though much about color rarity is regionally influenced, in general orange is extremely rare and white seems to be most common worldwide.

I once heard a collector say that "the rarest color of sea glass is the color I have found the least of." He was very accurate, in my opinion. That's because sea glass collecting is a very personal quest and how a collector fares has quite a lot to do with where he or she searches. For example, I know some beaches in the Caribbean where more turquoise sea glass can be found than cornflower blue, but on the rarity chart, true turquoise lands much higher up on the overall global scarcity scale because it is rarely found in every other locale on the planet.

The pieces represented in the chart are all genuine, ocean tumbled sea glass, most of them Pacific Ocean pieces. The black glass image features pieces that were found along England's North Sea. I used the English pieces because we do not have much black glass in the Pacific Northwest.

Sea Glass
Color Rarity

Orange

Red

Yellow

Turquoise

Pink

Black

Teal Green

Grey

UV Lime

Cornflower

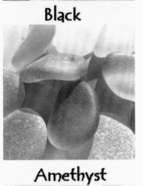

Amethyst

Honey Amber

Cobalt Blue

Olive Green

Aqua Blue

Seafoam Green

Emerald Green

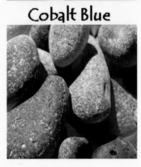

Brown

White

a few words about red sea glass…

Over the past decade, much ado has been given to red sea glass. Almost every time I work at a show, a dozen or more shoppers make a point of telling *me* that they have been informed that red sea glass is rare. The enjoyable part of the discussion is conversing with them about why it is rare.

What is so special about red sea glass? It is said that red is the most coveted of the many sea glass colors. Where does it originate? Why is it so rare? And can it be found anymore? A handful of smooth, rare red gems. After years of studying thousands of pieces, both from our collection and hundreds of other collectors' compilations from across the globe, it's a proven fact that red is one of the most difficult colors of sea glass to find. It's been said to take a lifetime of hunting to find just one piece. Red glass was not mass produced much in the United States due to the expense of the colorants needed to create pigment. The metal that helps create the color in the glass is from gold! This explains why red sea glass is so rare. If you do have a piece of red sea glass, it can originate from such items as signal lanterns and car and boat light lenses that were made from glass prior to about 1950.

On the left is a rare red nautical signal lantern on display at one of the International Beachcomber's Conferences. It won a blue ribbon for being such a unique find. These glass lens (not plastic, which show a more modern style) indicators helped aid boats and ships in navigating. It is very exceptional to find one with the glass still intact.

In my lifetime of sea glass collecting, I have seen only two pieces of sea glass that could be directly attributed to a buoy lens like this. One of them was found by my friend Linda's husband, Ben, along one of the Great Lakes. The other was found by a customer now sea glass friend Sue, who came to by my booth in New Jersey to have me identify her rather large chunk. I was able to pull up this image of the lantern from my website and tell her precisely what her piece came from.

Though I have quite a few pieces of red, most are tumbled so smooth that it is especially unique to find them with identifying marks like the ridge marks or faceting that is seen in lens glass. The reason why we have more than just a couple pieces in our collection is because ours is an older collection from decades of searching that was done long before the recent popularity of sea glass. Today such quantities and quality is unheard of.

A Pacific Ocean red on a rocky beach, right where I found it at low tide. Notice the triangular pattern in the piece, this clearly identifies this oddity as reflector light lens piece.

historic collections versus modern collections

I was giving a lecture and slide show one Sunday at the end of a convention. I shared photos and stories from some of my collecting journeys around the Pacific Northwest and many examples of my rarest and most treasured finds. Toward the end of each of my lectures I always allow time for the attendees to raise their hands and ask questions.

After this particular address, one woman asked why was it that the collection I was showing in my slides was so much smoother and baubly than her collection. She seemed somewhat exasperated and shared that she collects every day in Chesapeake Bay and never, ever finds the colors or the "frosting" that I find.

I tried to ease her concern and assure her that my collection probably looked quite a bit like hers. I shared with her that the pieces in my show represented less than 1 percent of my collection and that they were the best of the best samples of smooth, rounded, well-frosted "baubles" and some of the most special colors. I gave the example that for every five hundred rough-edged, plain white shards, I might have one smooth, blue teardrop-shaped beauty.

I also shared with her that the collection in my photos was a very old collection as could be determined by some of the unique color represented and that I have a lifespan of beach walking behind me. Older collections gathered piece by piece over many miles and years and hikes will have the more complete representation of historic pieces.

These rare turquoise, aqua, and cornflower blue sea glass smoothies have been tumbling along a rocky shoreline for decades. The longer a piece of sea glass remains in the elements, the smoother and more rounded it is likely to become.

I strongly believe that there are some older men and women out there who were collecting sixty and seventy years ago when they were children and have hung onto their collections. Their pieces might be sitting in old pickling or milk jugs deep in their basements or garages. These are the collectors whom I would love to find!

Almost weekly, I get phone calls from collectors who aren't as in love with the experience of collecting as they are with the experience of amassing fresh glass from more contemporary dumping. They'll phone and say they have "a lot" and want to know if I would be interested in buying their collection.

I make a point of directing them to some information about color and color rarity so that they can understand how glass color and specifically bottle color has evolved over time. The more current collections are going to have white, brown, and green glass and the older collections will show some blues, some pastels (including depression era pieces), and possibly even some reds and oranges.

The more modern collection may also be less smooth and frosty than an older collection due to its lack of time tumbling in oceans and being conditioned by the elements.

Though this common amber brown sea glass flat is a yummy golden color in the sunlight, it represents one of the most common colors of sea glass found today. This piece also shows that it has not been tumbling as long as some of its more rounded or oval-shaped colleagues.

genuine versus artificial

There was a turning point a few years back in the popularity of sea glass. It coincided with the growth-years of the Internet and especially with the creation of eBay. Though I now know that there are people like me who have been beachcombers all their lives and have collected for decades, I began to notice a new interest and a new population who had fallen in love with sea glass and its romantic and alluring side. Many of these new followers were landlocked collectors who had seen an article or found an online website describing sea glass and what it was.

Within a couple of years, the search-terms *sea glass* and *seaglass* turned up a fresh handful of sites and eBay listings. The popularity and worldwide appeal of sea glass grew quickly. Prior to this, when I was a child, I only knew of one publication, a poem, that even referred to the idea of sea glass.

> *"But, coasting that lone island round,*
> *Among the slippery kelp I found*
> *A little oval glass that lay"*
> —From "The Mermaid's Glass" by Henry Beers, 1917

During this growth spurt, I read some of the first articles about my very own collecting friends in major US and world-known publications. During the same couple of summers that Sharon in Hawaii, Richard in the Chesapeake Bay, and Lisa from Maine were featured, I was doing a radio show for a Toronto station and a story for the *Seattle Times*.

imposters

The moment sea glass collecting took the spotlight, its value as a commodity to be found, bought, and sold increased dramatically. And just like any collectable that has garnered sudden attention, the adoration caused a bit of a "run" on buying and selling.

And when any true item becomes of value, a faux or artificial product is not far behind. Man-made "frosted glass" showed up on the modern market. This mechanically created glass is *not* sea glass

Real sea glass is glass that has spent time and a unique journey at sea. It has stood the test of time and tide, often for decades. This is what gives genuine, beachcombed sea glass its value and significance. The process of mimicking the forces of nature cannot exactly be duplicated.

Genuine, authentic sea glass is glass (a bottle, a dish, an old window pane) that was once discarded, unwanted, and tossed out to sea as refuse. It may have found its way to the shoreline after being thrown overboard from a ship. It may have been barged out for dumping by a cargo ship. Or it may have been pushed off the edge of a sea shore town's landfill bluff.

No matter how the glass reached the ocean, we are finding that years and decades later, it is turning up along our beaches as gems with smooth edges and a frosty surface.

A truly mature piece can be so well-rounded and without blemish that it looks more like a marble than a sharp-edged shard.

Sea glass and beach glass is found on beaches all over the world. Some of it can be centuries old but most of what is found is most likely one hundred years old or younger.

We've discussed the reasons behind this earlier in the book, but it bears repeating: mass production of bottles began in the early 1900s. That's when glassware became much more common in the average household and was thrown out after being broken or unwanted.

It takes many years out in a natural body of water for a piece of glass to become smoothed, softened, and frosty. That frosty pitted surface is what many sea glass hunters and collectors admire. It *cannot* be mimicked exactly by a mechanized process or by a chemical bath. But there are some who have tried to create imitation sea glass.

To the purist, the historian, and the archaeologist in all of us, this has become an important issue.

My colleagues and I combined our genuine pieces from around the country and created a photo still life for educational purposes. Our pieces were arranged together on a table and then one member purchased some artificial "sea glass" and posed it next to the pieces of genuine sea glass. The groupings were photographed and posted on the organization's website.

The photo on top shows some of my genuine, ocean tumbled sea glass in the top left triangle of the frame. The more angular, blocky pieces in the bottom right triangle of the image are machine-made, frosted pieces of glass.

If you're a sea glass enthusiast, it is highly likely that you've seen this photo on the Internet and throughout many random blogs. Permission has been allowed for hundreds of members of the North American Sea Glass Association to use it for educational purposes.

However, it has been interesting to see the photo used on websites by unknowing users who believe it to be all genuine sea glass and with no explanation that it's a comparison image that holds both real and fake sea glass in it.

The explanation photo here shows that the pieces in the left of the image are genuine sea glass that are from the ocean and that the pieces on the right are from a manufactured process.

how to tell the difference

Every time my company has a booth and we display our product at an art show, we get the question: "Well, can't someone just make this stuff?"

My answer is always a very calm yet clear explanation that if it's man-made by a machine or a chemical process, it isn't really sea glass. Sea glass pieces are historic bottle and (usually) tableware glass that's been on a journey along the ocean for decades. To the purist, there is a big difference between the two types of glass because they are most interested in the journey the piece has been on. . . .

In 2004, when I was volunteering many hours with the North American Sea Glass Association, several of us decided that educating the public on the value and history of genuine sea glass was an important service we could offer. We decided to create a website page that listed the signs to look for in differentiating a piece of real, ocean-tumbled glass from a piece of glass that's been frosted to look like sea glass.

In fact, I still will not call "fake" sea glass, sea glass. I call it craft glass or frosted, machine-tumbled glass. I've always believed that if it did not come from the sea or a body of water, it should not carry the name "sea glass."

A team of us drew up a description as such:

genuine sea glass

- Originates from discarded bottles and tableware or glass from shipwrecks and household items lost in natural disasters
- Quantities of some colors are severely limited. Colors such as orange, red, yellow, cobalt blue, purple, true turquoise, "black," and Vaseline are very rare. Genuine sea glass in these colors is normally never sold in bulk.
- Sea glass is often hydrated and may have a "frosty" surface, appearing crystalline in structure. Hydration is a slow process where the lime and soda in glass is leached out by the constant contact with water, leaving variable pitting on the surface of the glass. The soda and lime can combine with other elements to form tiny crystals in the surface of the glass. Many good specimens will sparkle in the light. It

is impossible to duplicate this process without actually allowing nature to take its course over several years.

- Small C-shaped patterns may emerge on the surface of the beach sea glass and small hair line cracks may develop on some pieces.
- Natural tumbling usually takes place on uneven, rocky shores where a piece of sea glass got stuck with a portion of it still exposed. This process frequently produces shards that are triangular shaped, and yet in some areas, such as sandy beaches, the tumbling may be very even, making them well rounded and nearly uniform in shape.
- One of the best ways to discern if it's real sea glass is by realizing that much of it would originate from historic bottle glass. These pieces can be identified by bottle neck lines, rims, handles, and bottoms. Curvature is also a clear sign that it may have originated from bottle glass. True sea glass may also show patterns, lettering, and numbers. These features point to older tableware and bottles that had raised labeling—a practice that is rarely or no longer used and doesn't show much in manufactured, machine-made frosted glass.
- Sea glass continues to go up in price as supplies dwindle (littering is discouraged) and more and more people become collectors.

This cobalt blue sea glass triangle is naturally tumbled for decades. It clearly shows nice frosting and the angles and edges are well rounded off. The pitting is finite but still visible.

artificial

- Originating from either a factory, workshop, or rock tumbler (in rare occurrences people bring premature sea glass home to finish it off in a rock tumbler). Craft glass may be made from sheets of glass cut up and tossed into a rock tumbler or acid bath. Craft glass can also come from recycled glass bottles. Some who are a bit more particular will actually seek out old bottles to then turn into tumbled, craft glass.

- Nearly all colors are readily available in quantity and pricing between colors shows a consistency that does not reflect color rarity. Since one does not cost more than the other to produce, it's a sure sign of artificial sea glass.

- The most intricate factor that defines real sea glass from faux is the frosted, patina surface. To duplicate the hydration process that genuine beach sea glass undergoes, many manufacturers will etch the glass in an acid bath after tumbling it. Improperly rinsed, the glass may still contain some acid residue, which can be toxic. Some large craft stores that carry tumbled craft glass caution you against using it in your aquarium and to avoid excessive handling. This type of tumbled glass is often used in the floral industry in vases to support flowers.

- Etched glass has a satiny appearance and will be very uniform in its finish. It will be devoid of any small C-shaped patterns on the surface (which may occur on genuine beach sea glass).

- Tumbled glass is often quite rough on the edges. If it is well worn, the batch will usually show smaller pieces and each will be similar in surface patina "frosting."

- Many times tumbled craft glass comes in large, chunky, amorphous shapes, and sometimes it comes as nearly uniform squares and triangles. The packages that I have seen in the art stores show a much thicker glass than was used a hundred years ago in bottle and pane glass.

- Tumbled craft glass has a market and the differences are easy to see up close. It pays to be informed and ask questions.

This cobalt blue shard shows us what artificial, machine altered, frosted glass looks like. Though there are many ways to create frosted glass that tries to mimic sea glass, this is not sea glass in any way. This glass piece has not been on any kind of journey at sea and it is likely not very old. It did not originate from an antique medicine bottle or American tableware. It is simply a decorative piece of glass with an evenly muted patina.

always ask questions, always demand answers

The issue of genuine versus artificial sea glass has even led me to create an educational page on my website that informs the sea glass buyer to be wary and know the difference and that the difference between the two is important.

The modern buyer who cares that he or she is purchasing the real deal would be wise to know that today there are many sources to purchase sea glass from but that fraud does exist. The better educated the consumer is, the better the chance of finding the real deal.

The buyer should always ask before buying. Believe it or not, there are still some sellers who will advertise their glass as sea glass even though it has never been on a beach.

Asking if it is "tumbled" or genuine is not enough because the ocean tumbles glass too. Terminology is important. The buyer should ask several key and clear questions:

1. Are you the collector and did you find this piece yourself?
2. If not, can you tell me who and where you may have acquired it from?
3. If you're comfortable, can you tell me what general body of water the piece was beachcombed from?
4. Has this glass been mechanically altered, etched, sandblasted, or acid bathed in any way?

Most of the time the buyer will get a fairly honest and accurate answer. And sometimes the buyer's questions won't be answered at all or the answers will be false. If the buyer has a good handle on the signs of faux sea glass or if they have a connection to one of the more professional sea glass experts, they might be able to define if it's real or fake.

Many times the seller is just plain unknowledgeable.

In 2009, I found a craft site selling frosted glass created in a tumbling machine and then packaged. The website advertised the blue "sea glass" by the pound. But the photo being used by the seller was a photo of a pile of *my* Pacific Ocean sea glass—*a photo that I had taken myself!* The business's webmaster had found the picture on my online blog, copied it, and used my photo to advertise their product!

In this case, the seller and web designer were both uneducated and likely didn't have a very good photo of their "artificial sea glass" so they found one online to use.

Some sellers are simply uneducated in the differences between genuine sea glass and artificially tumbled craft glass. Some try to pass off craft glass as beach glass, and there are plenty who sell the genuine article and spend hours searching for it.

If the picture in the listing or the bag of sea glass shows very uniform color with no variance in the hue from piece to piece, odds are that it is tumbled craft glass. Blurry, poorly-lit pictures are also a red flag.

Pricing is not a reliable indicator of authenticity as some sellers price their batches, lots, and singles differently than others based upon what they have been able to find in their particular locale, and prices can vary depending upon demand. That being said, if you see a pound of a rare color like red or orange or any of the blues advertised for under $20, odds are that it is not genuine, natural, beach sea glass.

identification of most common finds

On a fairly regular basis, my little company gets phone calls from collectors who tell us they have "a lot" of sea glass and they've heard it can be valuable. They usually want information on selling it to a company like mine and are willing to meet me with their Tupperware bins or garbage bags full of their finds. Ninety percent of the time I have to let the collector down gently by inquiring about the color variety they may have assembled. At this point in the conversation, it usually gets more serious and most collectors tell me they have some beautiful green in addition to mostly browns and whites.

I inform them that those three colors make up about 90 percent of the collected sea glass out there and that, at this time, we are not interested in those shades. What I usually don't say is that if they have those three colors then my speculation is that the condition of the pieces lacks top qualification also. I once found antique bottle varieties in a tiny and dark antique store on an island in southern Canada. They seemed so forgotten and forlorn to me. But they added a light character to the equally ramshackled old window pane on which they were displayed. That windowsill full of bottles showed us that for decades, white (or clear), brown, and green were historically the most easily manufactured bottle colors.

Amber or brown sea glass is one of the most common sea glass colors. Beer, bleach, and sauces like Worcestershire have been bottled in dark brown bottles for years.

Though fairly common in color, this creamy brown bottle lip piece is more desirable as a collector's item because of its flawless condition, perfect frosting, and its visibility of thread lines.

Frosty, white sea glass can be very pretty and surprisingly pristine. White sea glass was originally clear glass with a surface veneer that has become cloudy white with pitting and "sanding" from years in the surf.

Many liquor bottles were created in this color in the early 1900s.

Over the past ten to twenty years, it has been a fairly painless feat to find an oceanside dumpsite and gather the newer, more common mass-produced bottle colors.

Deliciously rich in color, these emerald green pieces look like frosty gemstones in the sunlight.

identification of unique pieces and markings

Finding any piece of sea glass whether it's a common brown piece or something amazing and rare can always be special. But some kinds of finds are considered more memorable and more exceptional than others. In this section, we look in more depth at some special pieces and discuss where they were found and why they are considered so unique.

hey?!

Though it appears to say "HEY" the lettering on this uber rare, orange sea glass flat originates from the word CHEVROLET in all uppercase letters.

The piece really comes from a vintage, Pacific found, pre-1950, Chevrolet signal lens. The size of the piece which is about 5/8 inches tall, combined with its 1/8-inch thickness and the fact that it was not curved tells us it is a lens plate. Orange Crush, Fanta, even Boylan Seltzer bottles (1891) are good guesses, but because their bottle glass is curved, we ruled them out.

beautiful bonfire

One of the most often asked questions we get when doing lectures and sea glass identification is: "What is this multicolored piece?"

Very often it is a piece of glass with multiple colors in it that seem to "melt" right into each other. This very rare lemon yellow and frosty white melt is a good example. It is Bonfire Glass.

Years ago when people used to pitch their garbage onto the shore, they often lit a bonfire or incinerating pile. Here, there must have been two bottles or glass vessels that, when heated in the fire, melted together. The chunk melts to the sand and sometimes to other debris. After the fire has cooled and gone out, all that's left are pieces of glass to tumble in the surf. This now two-colored bonfire melt has been sanded and tumbled smooth with the passage of time.

Learn more about Bonfire Glass on page 72.

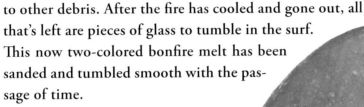

fish on!

This vibrant limey green, uranium dioxide sea glass piece was found with a thatch pattern on all three sides. (It's a triangular piece.) The ocean lover sees a fish, of course! The piece likely originates from a "Vaseline glass" tableware dish. circa 1910–1950. It glows vibrantly under a black light, as you can see.

Shown photographed under a black light, this vibrant, three-sided sea glass gem glows brightly due to a trace of uranium dioxide added to the mix during the glass manufacturing process to give it a yellow/green color.

Our traveling display shows some of the rare, limey green and yellow UV glass in the foreground. Though most collectors know that lime-colored pieces may contain uranium dioxide, not everyone is aware that yellow, red, and orange sea glass can have trace amounts too.

cobalt blue

These cobalt and soft blue bottle parts are from Nova Scotia's Atlantic shore. Cobalt blue is probably the most sought after color of sea glass, probably because it reminds people of the deep blue sea. It is easy to tell that these are bottle parts because of the neck rims and threads.

eggs

Rich and deep with color, this cluster of "egg"-shaped pieces are all very uniform in size and conditioning. Each of these pieces was found on a very rocky shore with some of the best wave action in Cornwall, England. It is possible that they've each been tumbling for one hundred to two hundred years. Notice the difference between the batch of blues below and this pile of smooth eggs. It is possible that all the pieces have been ocean tumbling for the same length of time. Some shores just tumble glass with a different patina outcome than the other.

Turquois blue sea glass: by our research, this is the rarest blue on the planet. Sometimes called *electric blue*, it can originate from late 1800s soda bottle glass and historic tableware pieces and it is very rare to find.

flat tire

This aqua blue bottle stopper flat top is well frosted, thick, and rounded (though it has a "flat" at the bottom). The wheel of beautiful sea glass originates from an old glass bottle stopper. The stem has long been shaved off from decades tumbling at sea and what's left is this nearly round top. It probably was used to cap the top of a vintage sauce or seltzer bottle.

pink eye

This multi-colored pink sea glass gem was also found along England's North Sea. Its bright pink color and the very unique symmetry makes this amazing piece look like an eyeball.

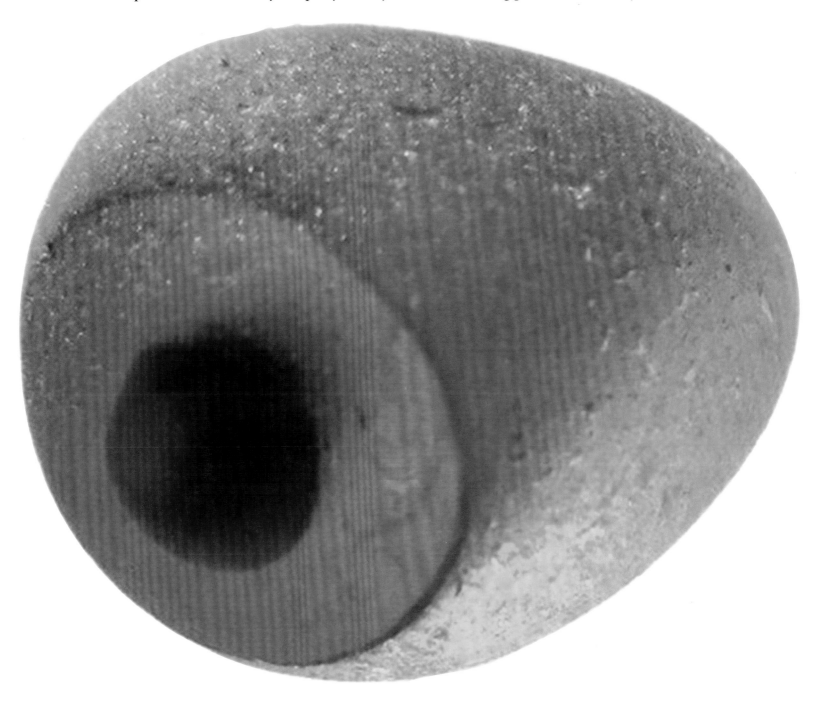

multis

Delicious swirled colors show through these sea glass "multis" found along England's North Sea. Some glass factories produced every color imaginable during the manufacturing process. Here multiple colors are mixed together when the glass was in its molten stage.

red and orange facets

Patterned and faceted red and graduated orange sea glass pieces that have all been Pacific Ocean tumbled. Most of these can be easily defined as being from pre-1950 vehicle signal light lens glass. Notice the bumps, lines, and faceting on the surface of the glass. This faceting helped the lens to reflect light more effectively.

rare red stopper

This teeny, tiny bottle stopper is probably one of our top five rarest pieces. This piece measures only about half an inch long. It was found on a warm California beach more than fifteen years ago. My best guess is that it once capped a very delicate bottle of perfume or oil.

Fishing floats: as I hiked along the lava flows of
Mt. Kilauea on Hawaii's shore, I noticed some sea
glass pieces would gather in the coral patches nestled
amongst lava beds. I was able to identify many pieces
as being from the Asian glass fishing floats that travel
a lifetime across the seas to break and tumble along the
rugged shore.

Candy. This frosty pink bottle stem sits on a bed of soft Florida sand. Because there is no bend or horizontal curvature to this piece, I assume it is a stem and not a rim from the bottom (or top) of an early 1900s candy dish or tableware piece.

Banana colored sea glass. Probably from a Depression
Era tableware piece, this golden yellow color accounts
for only about one out of every eight thousand pieces
in our collection. The near oval shape makes it perfect for
setting into a nice ring or piece of jewelry.

"crackle" sea glass

The type of decorative glass in which cracks were sealed into an art glass piece to give it a textured look is called *crackled glass*. It was popular in the early 1930s. But this piece isn't from crackle glass. This is sea glass that has been weathered this way with hydration, pitting, and fractures. Notice the tiny grains of sand in the crevices.

rare finds

Rarest of rare; these patterns, facets, and thread markings help us to identify these orange (and one bright yellow) sea glass pieces as from tail light and lens glass.

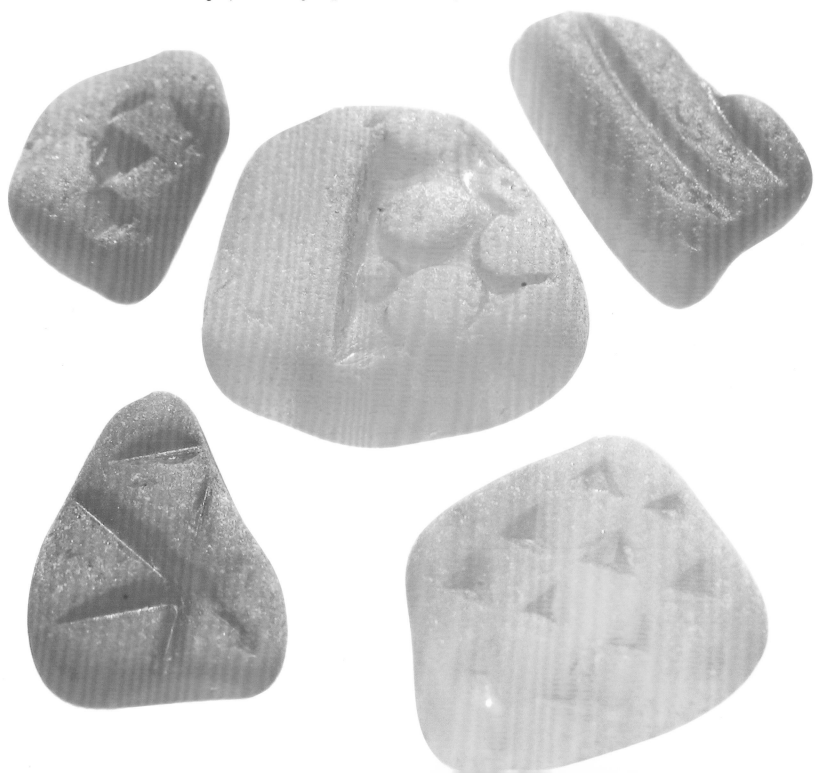

Nubbin. Limey-er than most seafoam green sea glass (which is more common), this adorable "nubbin" is perfectly conditioned and sanded smooth and frosty. It is about the thickness of a pencil and about half an inch tall. It likely originates from a turn of the century apothecary stopper with a top that has been shaved or tumbled off.

the tiniest gems

There are thousands of tiny pieces in this pile of sea glass. Some of the world's rockiest shores with the more tumultuous wave action are tumbling sea glass into the tiniest sizes that we've ever seen. We are watching this phenomenon with great interest. Some pieces being found on some beaches currently are smaller than the head of a pin. We are noticing this wonder on one beach in Kauai, Hawaii, for example. In the 1990s, it was common to find pieces there in the half-inch range and larger, but compared to the pieces we're finding today, those sizes would be considered gigantic. Recently we gave away one thousand free pieces of tiny sea glass in an online contest we were running. Several artists wrote to us and shared their visions of creating mosaics and art with the pieces should they win. I believe that is because when we think of sea glass, we still think of chunks that are in the one-inch or larger range.

flawless finds

As stated earlier, there are many factors that contribute to "tiny" sea glass. But here we're showing some rare, gargantuan finds that can be categorized as flawless. Here is a 2.5-inch lemon-colored, thick bowl bottom. One of the reasons it is so large is because, though it was found in a rocky environment (which helps create the fabulous frosting the item is showing), it was found more than fifteen years ago. Had it spent the last fifteen years in the surf, it is highly likely it would not be the impressive size it is now.

This brightest, glowing yellow sea glass marble in the left-hand corner is showing its years of being submerged in a sandy environment. The pitting on the surface of the marble has absorbed some shoreline sediment. This marble is extra special because it contains trace amounts of uranium dioxide inside. Uranium dioxide was used as a colorant in glassware in the early 1900s. It was banned in 1940.

At right, you can see the same marble as seen under a black light, which causes it to glow brightly. Shining a black light on your sea glass will reveal if you have any pieces with uranium compound inside of them.

Oooooh! While sorting for some earring-sized sea glass, we found a real surprise in one of our baskets of beautiful, multi-colored pieces. It's a gray/white with a red heart splash of color, found along England's North Sea.

This thicker, larger than we *ever* find on the West Coast, turquoise blue sea glass piece likely originates from a seltzer bottle lip. Photographed here on a rugged northwest beach.

bottle stoppers and buttons

The "sun lavender" bottle stopper on the right is the only one like it in our collection. The sun-"cooked" color indicates that it comes from older (pre-1906) glass that has colored a soft purple due to the sun's UV ray exposure. Though the color is uncommon, especially along the west coast of the United States, this style and size of bottle stopper is probably the variety that is found most often. The top of this stopper is slightly larger than the surface of a US quarter.

This section includes information about both unique bottle stoppers and sea glass buttons, mainly because they are both exactly that—very unique finds of which I possess few. I have one of the world's most complete and rare collections and yet I still don't have a large assimilation of either bottle stoppers or buttons.

Glass bottle stoppers were an effective way to seal a bottle when cork or screw tops were not used. Finding an intact sea glass bottle stopper is considered one of the sea glass collector's greatest achievements.

Finding a part of a glass bottle stopper on the beach is also quite a highlight. The part that a collector might find more often than a complete, unbroken stopper is the *shank*— the piece of the stopper that fits into the neck of the bottle or decanter. We often call this piece the *stem*.

The *finial* is the other bottle stopper piece that one might find tumbling along the shore. The finial is the top part of the bottle stopper that's made for grabbing. We often call this piece the *bottle stopper top*.

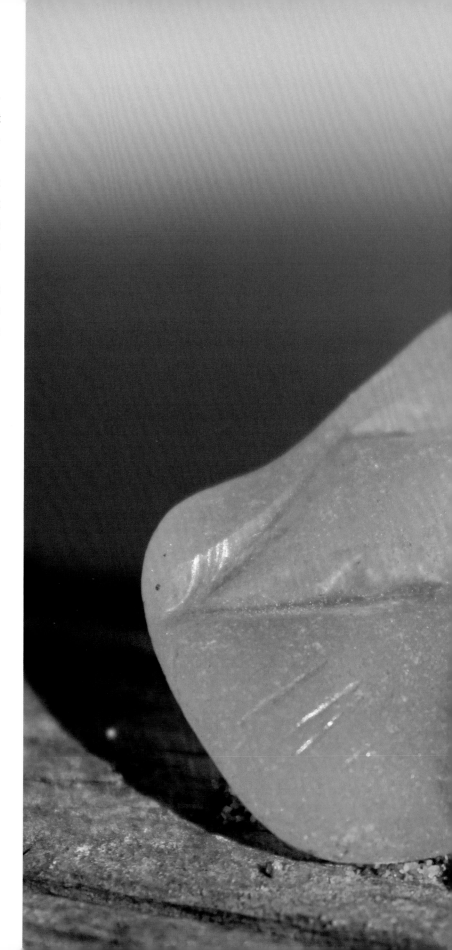

This clear, elongated white sea glass stopper was entered into—and won—the Bottle Stopper category at a recent contest I judged. The piece was not fully frosted; however, it was very delicate, which makes it even more incredible that it was intact.

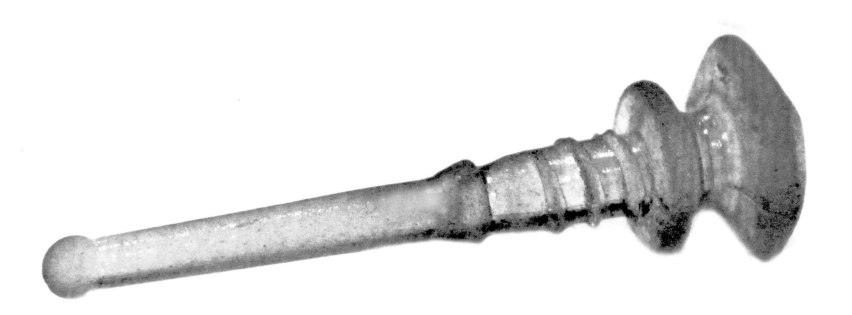

Sea glass buttons are also very uncommon. We don't have many antique buttons in the West Coast Sea Glass collection. One of our oldest, olive green buttons was set into a custom-made ring.

The only other sea glass button we have is an unbelievable, tiny cherry red. I suppose if we must only have one button, a red one is especially nice to have.

the lost bottle stopper

The two main kinds of glass bottle stoppers sea glass hunters search for are the rare, colorful decorative stoppers and the more practical, commonly used glass stoppers.

Which brings me to a heart wrenching story about the time I lost one of my most prized sea glass bottle stoppers: I was in warm, Northern California one fall while on a trip to organize a sea glass event. *Coastal Living* magazine had heard about the festival and had sent a writer and photographer to meet me and several other collectors for an interview and beach shoot with our pieces.

I was walking barefoot in the surf, holding a handful of pastel-colored rarities while posing for a photo. The shadows were long, so the photographer asked me to turn and face her with my back to the waves; something a beachcomber should never do! As expected, a larger than ordinary wave smacked me in the back. Startled, I jumped forward and threw my hand upward, accidentally loosening my grip on the sea glass pieces. Only one snuck through my fingers but, unfortunately, it was my true treasure. It was a tangerine-pink bottle stopper from the early 1900s—probably something that capped a decorative perfume bottle or carafe.

I turned and watched the piece land several feet behind me in the deeper surf just as another, larger wave broke over the top of it. Soon, several more frothy waves were converging in and swiftly running back out to sea. I ran in and shouted for assistance. "Help me find my piece! It was the pink bottle stopper!"

We looked for several minutes with no luck. The photographer and writer were patient even though we were on a timeline and had another appointment to get to.

I ended up shrugging it off and tried to find comfort in the thought that some very fortunate wanderer would soon come across the piece of a lifetime.

sea glass marbles

The true, well-pitted, blue swirl sea glass marble at the right was found after decades of natural tumbling along a Pacific Ocean shore. This marble is one of the most beautiful and well-pitted specimens I have ever found. It shows perfect frosting from years at sea and it has also kept its shape and is swirled with color. It is difficult to tell exactly where it originated, but some theories on how marbles like this one end up on the beach are below.

Like other bottle glass or discarded pottery or toy parts, occasionally a true sea glass marble piece can be found on the beach. It is considered very rare, even to the seasoned sea glass collector, to find a marble or piece of a marble along the shore. All marbles I know of have been picked up from along Pacific Ocean beaches. The rugged and rocky shores of the Pacific tumble our sea glass quite nicely.

What were they used for originally and why, so many decades later, can they sometimes be found rolling around the beach? How did they get there?

how do marbles end up on the beach?

There are several theories describing why marbles occasionally wash up on the world's beaches.

Reason #1

In the late 1800s, an inventor named Hiram Codd designed a glass bottle that used a marble as the stopper. The Japanese glass Ramune bottle was also sealed with a marble. These two bottle styles were used in the United States and around the world and likely account for many of the beach marbles that have been found (and can still be found, though rarely) along the shore. When a bottle was discarded, often into the sea, the bottle would break against the rocky shore and the marble might stay intact and tumble there for years.

Reason #2

Years ago, many ships were loaded with heavy items to help provide ballast. Marbles provided this weight inexpensively and effectively. In the Puget Sound, where the tides move fast and the inlets can be narrow, ballast is key. It reminds me of the white water rafting trips my family goes on down remote Hells Canyon in Idaho's backcountry. The heavier, more weighted-down boats fare much better in the turbulent rapids than the lighter rafts. Likewise, ships along an ocean's rough shore may have needed this weight to help with navigability.

A sea glass collecting friend of mine, Stephanie, used to live and work in St. Thomas Virgin Island and was an avid sea glass marble collector. She has written me multiple times with a story of how, one blessed day, she found more than just her usual one or two sea glass marbles. She was trying to solve the mystery of why the marbles ended up there on her beach.

She was hiking along a shore that was lined with steep, sandy cliffs. One afternoon she had found one or two marbles up higher on the beach bank. Then she found another that led her up, away from the water's edge to another, until she found herself staring directly into the cliff face. With nothing but her bare hands, she decided to dig into the clay-like cliff's side. In just a couple scoops of sand, she said, several marbles came tumbling down, right out of the wall itself at about waist height!

She did some research and believes that they may have been poured out there years and years before she came to that beach. She had heard stories of the rum runner during the early 1900s that carried barrels on sloops back and forth throughout the Caribbean to fill with alcohol. She shared stories of how the barrels were oftentimes filled with heavy items prior to their pickup so that the ships had heavy ballast. Marbles would have provided a heavy, yet fairly inexpensive cargo that could easily be dumped along a shore once the ship arrived for a pickup of illicit rum.

Whether a vessel dumps its load or the ballast is tossed about in a storm, the cargo could easily be smashed upon the rocks. The boxes or, in this story, barrels, of ballast marbles would surely be lost to sea only to wash up on shore decades and centuries later.

A color collection of Caribbean and Pacific found sea glass marbles.

Reason #3

Decades ago, young children often played with sling shots and may have used marbles for ammunition. The beach made a great place for target practice. My father, who was born and raised along the Atlantic coast—specifically along the Chesapeake Bay in Virginia—told stories of himself as a child playing games with his siblings and friends by floating a "moving target" piece of driftwood offshore and then shooting their marbles out into the water toward the target. Some seagulls often became moving targets too. The resulting marbles, which landed just offshore, one day washed beachward.

Reason #4

Painters were known to often drop a handful of marbles into cans of paint to help mix the batch. When the paint was used up and the can was tossed into the city dump (oftentimes the dump was the sea or off a bluff), the salt water and ocean's natural biodegrading ability decomposed the paint can over the years. The marbles became what was left and each washed upon the shore until individually beachcombed.

If you can find it, look for photos of some of our rarest and oldest marbles in the August 2008 issue of *National Geographic*.

This high-quality, dusty blue swirl marble caught my eye as I was judging the 2013 North American Sea Glass Festival Shard of the Year Contest. The annual contest has a Sea Glass Marble category.

sea glass hearts

A once broken and sharp shard that has now been frosted smooth by the elements is always considered a treasure. The smoothing nuances of the ocean's waves, wind, and sand on a piece can never be exactly mimicked twice. So when a special shape emerges from the once hard and shiny glass, the moment you stumble upon it becomes even more momentous.

Shards of glass are most commonly found as triangle shapes. This has everything to do with the angles that discarded bottle and plate glass most often break into. If the piece doesn't get tumbled for too many years or decades or if it is tumbling along a shore that does not get a lot of wave or tidal action, it may stay fairly triangular throughout most of its life at sea. It is quite common to find triangular-shaped sea glass, especially if it's a piece that is fairly contemporary in its age.

Though a heart-shaped piece of sea glass is an extremely rare find (not to mention incredibly romantic), it really is just a more intricate relative of the triangle. Most heart-shaped pieces are never a perfect heart. Rather than originating from an equilateral triangle, most heart-shaped sea glass originate from what is called a *scalene triangle*. This is a triangle that has no sides of congruent length. If it has two congruent sides, it is called an *isosceles triangle*.

Most sea glass hearts are just that: non equal-sided triangles with a divot or pit in the middle of one of the sides.

A sea glass heart that is smooth, well frosted, and not fresh with sharp, angled edges can take decades for the ocean to perfect. They are not common at all and do have quite a collector following.

One factor that can make an already rare sea glass heart even more special is if it also is rare in color. The quintessentially romantic and rare sea glass heart is the orange or red heart. If it has lettering or faceting on it, those factors raise it to an even more exquisitely rare level.

Most sea glass hearts are saved, cherished, and gazed at for years. But, occasionally, the artists at the West Coast Sea Glass studios do choose to invest time and craftsmanship into setting the piece in a classic keepsake piece of jewelry. Some of the finest hearts in the most perfect shape are even made into heirloom Valentine's Day gifts or wedding jewelry.

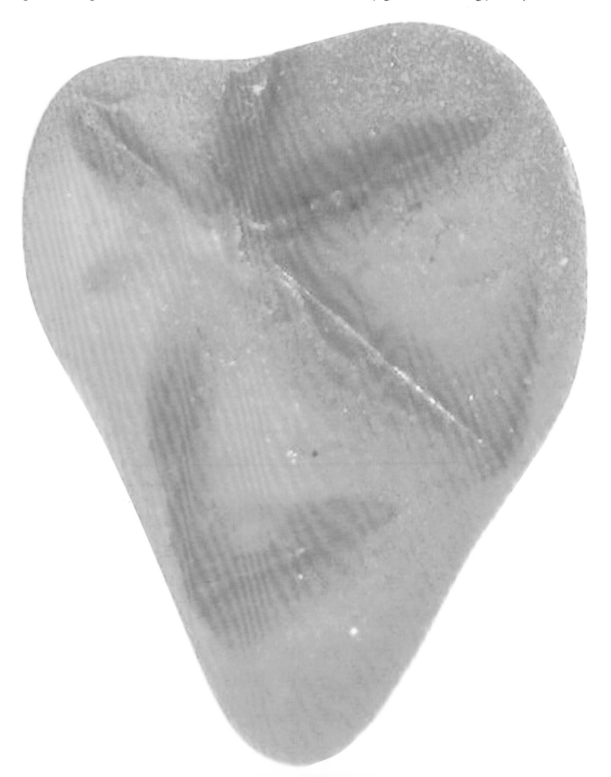

be mine?

On more than one occasion, I have had a customer phone us requesting to purchase a rare heart in a certain color, meant for a special gift or occasion.

One particular young man romantically planned to map out a beach walk with his beloved on the afternoon he would be proposing to her. Ahead of time he purchased a nickel-sized, cobalt blue heart. Then he trekked out to the place where he would propose. He dropped the heart along the shore—not too close to the tide's edge and not in a highly trafficked area so that someone else might serendipitously find it. Then he picked up his soon-to-be fiancée and took her on the beach walk and directly back to the place he had "planted" the heart. They walked slowly and searched for shells, beach rocks, and sea glass. Then ahead, in the sand it lay. He stayed a few steps behind her so that she could chance upon the piece. She gasped when she saw it, picked it up, and that is where he proposed.

It is my guesstimate that for every 3,000 to 5,000 pieces of sea glass we find, one might be a true, sea-smoothed heart.

This deep crimson sea glass heart is larger than most red pieces we've found. Its size enabled it to work perfectly in this romantic, heart-shaped, silver ring.

art glass finds

Glass blowers have been creating artistic glassware for hundreds of years.

Art glass has been around for hundreds and even thousands of years. We touched on this topic slightly on page 44 when we mentioned that the Chinese, Greeks, and Romans are credited with advancing the creation of functional glass pieces to a higher art form. Even the Venetian glassmakers in Italy, who were moved to the island of Murano, are credited with raising glass's artistic reputation worldwide.

Keep in mind that glass manufacturing moved to glass blowing with a tube, which moved to glass molding and glass staining. And throughout each movement and creating style, there is no doubt that the term *art* was applied loosely and often while the pieces were being created.

Though much of the early American cut glass and colored glass was created to serve a utilitarian purpose, the term *art glass* was attributed to bowls, salts, carafes, and other more decorated household items.

Today, when people use the term, they are mainly referring to glass that is made primarily with a decorative purpose.

Most of the true art glass seen around the United States is fairly modern when juxtaposed with the overall picture and timeline of glass manufacturing and following historic glass making from its inception. These pieces have been blown in studios and galleries specifically dedicated to an expressive art form where glass has been the medium of choice.

One of my first, up-close experiences with a true glass artist occurred when we shared spaces directly next to one another at one of the country's biggest art festivals. This was such a specialized art fair that every artist was juried in and required to set up "full shop" to demonstrate their craft during the entire show.

What a treat to be able to watch this focused individual weave and finesse and toil through the day on his craft. He had torches and tools and fire and shafts. And the colors! His booth was lined with rods of every shade and density as well as opacity.

Multi-colored Lundberg Studios art glass. These pieces with milky white opaque glass mixed with blue and green areas of color were found on a beach near a glass blowing studio.

where do art glass pieces come from?

Art glass pieces that have made their way to the ocean or body of water have arrived there by a couple means. Most of the time, if a beachcomber finds a piece of sea glass art glass, it is because they are collecting along a shore near where an art glass studio used to be located or, in some cases, still is. Glass blowers along beach towns and boardwalks historically disposed of their garbage in the sea, just like most all of humanity did at the turn of the century.

The other reason a sea glass collector might find a piece of old art glass is because someone once owned a piece of the art and it broke. The item's parts were discarded and subsequently made it to a shoreline landfill.

My beachcombing friend, Michelle, experienced this exact circumstance. She was beach combing on a long sand spit at the northern edge of the Olympic Peninsula called Ediz Hook. This locale has never been known for art glass finds. If a beachcomber does find pieces there, they are usually tiny, smaller than a half of a peanut, and white or green in color. But her piece was creamy and concave with layered black, yellow, and orange swirls. And it was large; about the size of a walnut.

She showed it to me and neither of us (both seasoned sea glassers) had seen a piece of art glass like that anywhere near that particular area. Familiarity with that location led me to the conclusion that the piece was something that had to have come from a ship's dumped load or some sort of refuse dumping. It did not come downstream from the river just east of there because there have never been any glass blowing studios in all the history of the area.

If a collector finds a piece because they are hunting near an American glass blowing studio, it is likely that they might find a second piece or a third. The concentration is proof that glass makers disposed into the water nearby.

When I was helping identify shards at the first sea glass festival in California, several attendees brought pieces to me that looked like intact, fairly modern paperweights. We also saw pieces of Murrina glass rods with bold flowers and stars and creamy whites, striped with greens, blues, and oranges. It was obvious there was an art studio nearby where these larger than usual and more modern than usual sea glass oddities were coming from.

Just down the road, there is a studio near the beach in Davenport, California, that's been in the art glass business for more than forty years. The Lundberg artist's glass creations have been found for years, tumbled nicely by the mighty Pacific. Glass has been known to wash up as sea glass near other studios in the Monterey Bay and Carmel.

A couple years later, the Travel Channel producers who worked with me on the sea glass show *Treasure Hunter* inquired about the Davenport site as a spot to find some more current pieces and a few colorful pieces were presented in the episode.

colors

Modern art glass is usually very colorful and we see a lot of opaque colors mixed in. And every color imaginable is used. I am not highly knowledgeable about the more modern (after 1950) art glass in the United States.

In fact, at one of our Pennsylvania events, I was announcing the winners of a rarity contest and my friend, Terri, held up one of the entries that was to be given an honorable mention; an obvious art glass piece. She was stumped, so she offered me the microphone and said, "Mary Beth, you tell them something about this piece." I hesitated. "It's art glass, modern art glass . . . that's why there are multiple colors in the piece." And that was all I was really able to say.

Since then, we have incorporated an actual art glass category into the contests and we've had the privilege to view some amazing sea glass that came from older art glass from other countries. See Anna's pieces on page 40.

what is a glass beach?

There are many *glass beaches* around the world. A glass beach is a shoreline location where garbage dumping was done and as a result, a higher than usual volume of glass has been left to tumble on the shore. Glass has proven to be one of the last items left still tumbling in the elements years after the practice of garbage dumping has subsided. Though the glass beach phenomenon is dwindling all over the world because dumping along our shores is practiced less, numerous coastal towns had or still have an old glass beach but many are no longer highly obvious.

Today, most of these beaches usually look like fairly untouched, natural beaches. If the beachcomber does their homework and researches where a town's landfill was or where shipping stops were, he or she might notice a few small twinkles of pieces showing. A glass beach is often below the base of a bluff or cliff, especially bluffs that have roadways or trails leading to the bluff's top edge. This would have been a road where cars or trucks may have backed in to dump their load. The beach walker on these otherwise naturally looking shorelines will have to pay close attention in order to notice small pebbles of smooth, rounded sea glass, pottery, or porcelain parts.

A glass beach can be located far away from the city or town, along a removed and far out-of-the-way shoreline, or it can be smack in the middle of the largest city's busiest marine port or bay. The next time you are walking along

This beautiful beach photo was sent to me in the summer of 2008. My sea glass collecting friend Julie walks beaches along Russia's Far Eastern shore. She tells of a unique rocky stretch with a couple of coves where parts of the shore are covered by various colors (mostly brown and white) sea glass. She tells of an old glass factory somewhere not too far away where waste was regularly washed away to the Sea of Japan and then tumbled ashore.

a bay that is busy with boat traffic, ferries, or even naval ships, look down to the tideline. Unfortunately you will most likely notice some garbage and small pieces of plastic flotsam but that is your clue that other small glass pieces might be found there too. Follow the beach down to the water line or up to the high tide line. If you find any concentration of broken glass you've likely found a place where humans have been dumping. The glass stays and tumbles and defines it as a glass beach.

While traveling in Canada, my family was walking on the boardwalk along a busy marine port, I looked out at the boats coming and going through the bay. And then I glanced down at the slight strip of pebbly sand that wove its way below the boardwalk and under the docks. There was broken glass everywhere I could see. Most of it remained rough edged as most of it was current, brown and white glass but still, it was glass. I wanted to skibble down the boulders to the beach and sift through the pieces but my safer self-intuition kicked in and I decided that another day when my young children weren't with me would be a better time.

Often glass beaches have several coves or separated areas of concentrated refuse. I happened upon one of these types of glass beaches one afternoon while walking with a beachcombing friend. We were in a rather remote part of Mexico's Baja Peninsula on our way to a more populated, better known beach spot where we presumed there might be some good sea glass. Not at our destination yet, we were engulfed in our conversation and not actually searching for glass. My friend spotted a soft yellow pottery shard, then

As you can see in the photos, the pieces are highly rusted from the remains of metals in the silt and sand.

another lime green one, then I found a coppery colored one. We had been walking for about a mile up to this point and hadn't found one piece of pottery. But here within a concentrated 100 feet space, we found two dozen pieces of pottery and glass. Looking up and out, we noticed we were in a slight bay at the base of the steeply sloping cliff above us. Here we discovered that we had come upon a smaller glass beach dumping area that was somewhat close to but definitely not right on top of where we expected the main dump area to be.

Usually glass beach is not a formal, proper name but it is a descriptive term and the term describes several beaches around the world.

For example, though I know of several glass beaches in the state of California, there is one there that has been formally named Glass Beach. Even the trail that leads to the more famous glass beach is now entitled, Glass Beach Trail. Today there is a sign where the road ends and the trail starts, pointing tourists to the water's edge. Almost every time a patron asks me whether I've been to Glass Beach, I make sure to mention that "many states have glass beaches. But I'm guessing you mean the Northern California one?"

Today, glass beaches also present us with a bit of a responsibility. Since they are known as former dumping locations, most sea glass pieces found there can really have quite a bit of history. I have even heard people liken the history of a sea glass piece's journey to something akin to beach archaeology. To many sea glass lovers, the age of a piece and its unique voyage over time is very important. Therefore glass beaches by definition can literally hold "time in a bottle."

The care that some coastal enthusiasts and historians take over such beaches is a very important part of the sea glass story. Never again will we see the kind of relics and unique pieces of the past on our beaches. The bottle glass, art glass, and tableware industries of the past 100–200 years are not likely to happen again and I'm sure that shoreline dumping will continue to cease around the globe as we become more ecologically conscious.

so what about the idea of glass "planting"?

It's not an unheard of notion. If you can believe it, some sea glass enthusiasts are concerned enough about beach glass depletion that they've considered, and are even practicing, glass dumping on beaches today. The most fervent of these collectors point to a desire for their children and grandchildren to have the sea glass collecting experience some day. It is true.

It can take decades for a piece of sharp, broken glass to become smooth from the ocean's tumbling power.

I remember my own, first conversation about this idea. It was the end of day one of the very first national sea glass festival that many of us had just finished organizing. The booths were buttoned up, the exhibits were covered and the convention hall doors were locked. I was on my way out to meet up with my dear sea glass friends at a California tiki bar. I left the exhibit hall and walked in the evening sun toward the outdoor restaurant. My friend and well-known sea glass author, Richard, caught up with me. We exhaled with relief and were pleased with an event well-done.

"After today's crowd, it's obvious that sea glass has really caught on. What's next for us?" he asked. We discussed.

"A lot of people asked me about what we're going to do when all the sea glass is gone," I said. "People are even sharing ideas of breaking bottles, hauling the pieces out off shore, and dumping them so that we have sea glass to collect forty to a hundred years from now."

Though glass behaves as a fairly inert ingredient along rocky shores and in a salt water environment, the two of us concluded that a practice like this seemed like a step backward after how far we've come as a generation in regards to ecology and shoreline dumping.

We talked further about how private, beach property owners might be more likely to attempt something like this. Maybe the organization should look into buying a remote island and starting such a project, for research purposes, of course, we supposed. At one serendipitous point that very summer, I met and befriended a realtor who actually had a private island, complete with pebbly shores and steep tides, listed as one of her properties. We discussed the possibility and curious, I researched the reasonably priced, one million dollar property. I smile at all of this now because I have always been one who seriously hesitates at the idea of planting sea glass.

why not "plant" sea glass?

What is very uncommon, as seen in my hand, is the flat, sharper-than-the-others big blue piece. Everything about this out of place find, points to the fact that it is likely a piece that was very recently pitched onto the beach. It doesn't hold the same historical background that the other pieces hold and it's not been on a lifetime journey like the others. If it had been here tumbling for the same amount of time (fifty-plus years) as the other pieces, it would be smaller, smoother, more rounded. It also would not have been found there. It would have been quickly gathered up by a beachcomber years before I came along.

The practice of glass planting is still debated. But the reality is that disposing on otherwise historic sites creates quite a jumble of confusion for the historian, the sea glass collector, and the archaeologist in each of us. Not to mention the tourist.

I had a customer who excitedly wrote us because he had found the piece of a lifetime; a perfectly round, flat on the bottom, semi oval piece. And, in addition, it was red! It was about two centimeters long and flawless with uniform frosting. He wanted me to give him some history of his unforgettable find and he told me what beach he found it on. The particular section of shoreline which he found it along only offers tiny, tiny (usually less than 1 cm) pebbles of sea glass and has done so for years. So happy for him, I asked him to send me a photo of it. Surprised, it was a much larger piece than the others and it closely resembled a very modern pebble of aquarium or floral arrangement glass to me; not something I have ever seen found on the beach he referenced.

One of these things is not like the other. Here we see many smooth, well-rounded, ocean tumbled sea glass tinies. I know for a fact that the softened brown, white, and olive green pieces are genuine, historic sea glass pieces from decades old bottle glass. I know well the beach where these were found and have a long history of hiking there. These same colors, with the same surface patina and same size have been found here at low tide for decades. Dumping has not been practiced along that particular shoreline since 1962.

These are the shiny, uniformly sized orbs that any modern day shopper can pick up at the craft store; usually the pieces are bagged by the color. I was honest with our customer and mentioned that I believed it to be a much more modern piece than what we usually see from historic dumping on that beach.

Interestingly, a couple weeks later I met a woman at an art show who stopped at my table and said, "Sea glass. My nephew, bless his heart. He makes sea glass!" I shared with her that this sea glass here is quite old and is something that is made by the ocean and is found on beaches. "Yes," she said. "He pours glass craft pebbles and flower vase filler glass onto the beach for the tourists. He recently put some red ones out there." She then shared where he did this and coincidentally, it happened to be the same beach where our customer found his pristine red orb.

In addition to the fact that it is considered dumping, which is illegal, and that it can be unsafe, and that it rearranges an ecosystem, this story gives us a perfect example of why glass dumping should be seriously reconsidered as a modern practice along our shores.

sea glass from around the world

I have had the opportunity to travel a bit around the United States and to a few other countries, oceans, and islands on the hunt for sea glass. I've catalogued some of our interesting pieces here.

Shown here is a very fun exhibit my coworkers and I created to display our unique sea glass pieces. It defines the kinds of pieces they are and identifies where (country and body of water) they were found.

NUMBERS WORDS PATTERNS, PACIFIC

NOVA SCOTIA, CANADA

, OH, PA

Exhibit On

VIRGIN ISL, CARIBBEAN

ITALY, MEDITERRANEAN SEA

PUERTO RICO, ATLANTIC

These particular pieces derive from a specific beach off England's shore. Sometimes a glass manufacturer would dump refuse out while the glass was still warm. The molten materials of differing colors have blended together to create these beautiful swirls. The items disposed of over the cliffs to beaches below would tumble smooth. From their previously unwanted state, they've been refined by the sea into a highly desirable and beautiful gems.

Here are some vibrant gems collected off a Puerto Rican beach along the Atlantic Ocean. Puerto Rico had many old houses with colorful window panes featuring a pressed starburst-like pattern in very bright colors; emerald, red, purple, yellow and even this vibrant, rare turquoise (left center). This is one of the only places in the world where we see this kind of sea glass wash up. This is a perfect example of how a piece can tell you where it was likely found and what it most likely originated from.

The colorful pieces at left were plucked off an island beach in the Caribbean Sea. I have been able to travel to the Caribbean a couple of times and I have been blessed to have had the time to comb some beaches. We seem to see much older glass in the Caribbean than we do from other shores around the world. The dark, dark "black" glass piece shows a color that we never see along the Pacific coast for example. To me this piece resembles old bottle glass, possibly from a whiskey or rum bottle bottom.

I showed this photo at a recent lecture I gave and I asked the audience where they thought this Italian glass may have originated. Several people guessed correctly: wine bottles.

The treasures at left were collected on a Virgin Island coast along the Caribbean Sea. A sea glass friend named Patience (a perfect name for a sea glass collector) has sent me many pieces she has found. I would guesstimate that 90 percent of what she finds are brown, white, and green bottle glass and fairly contemporary. One day she sent me a photo of a year of collecting, and the pieces covered the entire top of her poolside lounge chair.

Nova Scotia sea glass looks quite similar to sea glass found in the Pacific Northwest. I've deduced that this is, at least partially, due to being of similar latitude north of the equator, which means that each location experiences similar tides.

Notice the patterns, portions of words, and numbers on these pieces that were collected from Pacific Ocean beaches. If a large enough piece of sea glass is found that shows entire words or a distinguishable pattern, the original design or even the name of the specific pattern can be determined. For example, I am sure that the pink "swan" wing in the front left of the photo defines the piece as a Fenton art glass piece. The piece is large enough to tell me that it is most likely the left wing on a Fenton oval bowl.

The United States's Great Lakes have significant wave action that emulates an ocean's tide as you can see from these pieces collected in Michigan, Ohio, and Pennsylvania. Notice how these pieces possess a similar appearance to sea glass but because the pieces were collected from a lake, this glass is most often referred to as beach glass not sea glass.

Scotland has some of the best historic glass I have ever seen. Our collection of sea glass beachcombed off the North Sea contains unique bottle stoppers and many other exceptional finds.

These larger-than-we-find-along-the-Pacific pieces were collected from the northern sections of the Atlantic Ocean, just off Rhode Island, New York, and Massachusetts.

a grecian find

These pieces were found on a beach in Greece, off the Mediterranean Sea. A couple years ago we traveled to the Greek islands of Santorini, Paros, and Naxos and the Athens mainland. There we picked up a few beautiful pieces. Glass blowing began in Syria, far north of Greece, around the first century BC. Most Greek glass is likely to be much older than the sea glass found in the United States mainly because glass blowing has been practiced there longer than it has in younger countries.

As serious sea glass hunters, being anywhere near the water is always what it's all about, but being in the islands, walking through the teal blue waters, and enjoying the warmth of the Mediterranean climate made for a sea glass collector's paradise.

Several days were spent walking the beaches, hunting for sea glass, and enjoying the sunsets.

A sea glass friend who resides in Greece part of the year and whose husband is a yacht captain took us to a few spots on one of her favorite islands to sea glass collect.

We spent one morning on the island of Paros's beaches. We strolled through the waterfront town's streets and buildings and stepped into a few boutiques and art shops. The architecture avails itself to some very colorful and rich photo opportunities.

On the island of Santorini, steep cliffs and very narrow roads abound. One of the best ways to reach the most remote beaches is to rent scooters and zip around and down to the shore. Local inquiries and maps help too! Many a tour bus could not go down side streets like we could. More than once we stopped at a street-side shop for olive, caper, and grape hors d'oeuvres.

Hawaiian life

I've also had the opportunity to collect sea glass along several Hawaiian Islands. Though most days were devoted to sunning, snorkeling, and boogie boarding, I did hike about five miles one day in 86 degree heat to sea glass hunt on the big island.

With forty-plus mph offshore winds, the waves were high and the beaches were well churned up. This makes for prime conditions to wash fresh beach glass debris shoreward.

I hit a several mile stretch of beach that was virtually uninhabited. Though remnants of ancient homesteads built of stacked lava rock dotted the dry, barren coastline, no one lived there now. Below me along the shoreline, cove after cove was carved with black rock that thousands of years before was molten and had rushed downward to the sea from a volcanic eruption. Tide pools had formed in the crevices of the cooled lava rock and occasionally an oasis of white sand and coral rested there in a patch. A spot of color (usually soft teal blues and seafoam greens) would stand out against the pebbly background. Sea glass! Many of these pieces were clearly from ship refuse and bottles jars. It beckoned as a smoothed, rare treasure amidst this harsh environment.

Part Five

CARING FOR AND USING YOUR SEA GLASS

cleaning your glass

Since sea glass is found in an organic environment and has likely been making a home there for years and years, it will have bacteria, stains, and other biological material. Most sea glass has been in salt water surroundings from ten to possibly a few hundred years. Most of the bacteria on our sea glass cannot be seen and most of it is generally harmless. I know this because some of the salinity tests we give to our glass entail licking it. Yes, licking it.

The easiest way to clean your sea glass is to simply hold it under warm, running water and rub it clean with your fingers. If you'd like your piece to not only be visibly clean but also free from most bacteria, basic liquid dish soap can be rubbed on and then rinsed off the surface of the piece as well. Ninety-five percent of all sea glass should clean up this way.

Occasionally, a piece of sea glass can emerge from the shore with staining that is harder to wash off than with the simple rinse described above.

You may have noticed that a few of your pieces show some muted brown or reddish brown staining. Most of the browning that we see on sea glass derives from metal rust. This rust-colored stain is very often a telltale sign that your glass has been in an environment where other items in addition to glassware were discarded. Items that can create rust staining on your sea glass include metal cans, car parts, wire, junk iron, and more.

On numerous occasions I have viewed a person's collection and had to disappoint the collector by telling them their glass is not really rare yellow or mottled, rare orange like it may appear but that the piece is stained with rust. Many times I've been able to actually guess what beach they collect on because some better known "glass beaches" produce this discolored sea glass.

I recently met with a retired sailing friend of mine to view her collection and tell her about her pieces. Though she sails all over the world, the pieces she was showing me were reportedly from only one beach; a two-mile long location up in the Pacific Northwest. One of the bowls of glass she showed me was clearly different from its cleaner counterparts. The glass was flat, not very well tumbled, not well frosted, larger than the glass in the other containers, and dirty with orange streaking and blotches.

If you're a regular collector, you probably have quite a few sea glass pieces with seaweed or mud dried onto them. I call these rust stains "dump site stains" because very often this staining is common on sea glass that is found along coastal dump sites.

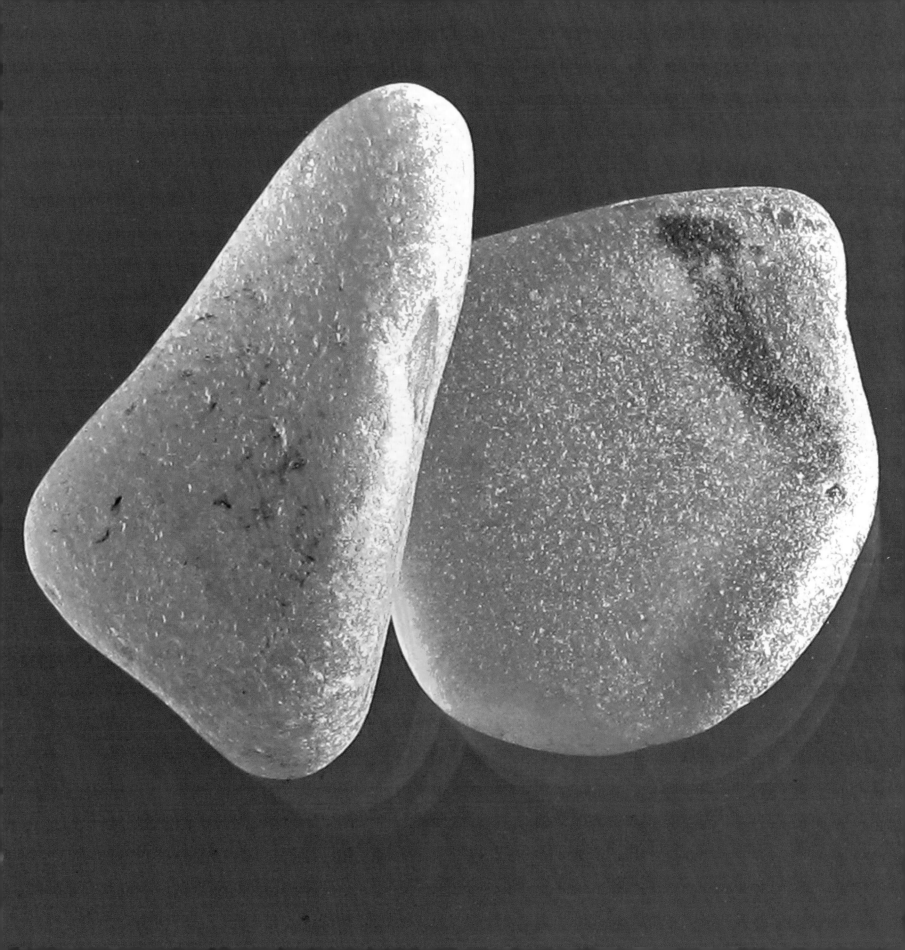

I sat up and mentioned that the bowl seemed like it might be from a different location than the other pieces in her collection. She shared with me that she'd been all around the world and had collected in various places. I then stated that the stained pieces reminded me of pieces I've seen from such landfill site beaches like Dead Horse Bay in New York. She reviewed the bowl I was combing through, then her mouth dropped open. "Oh, that's amazing. This batch *is* from there! We stopped there last fall and did some collecting."

The rust staining is very difficult to clean from sea glass mainly because sea glass has a very frosty, pitted surface. This surface tends to hold onto rust and dirt, which seems to saturate and sit inside all the microscopic pits and fissures on the surface of the sea glass. Even washing with warm water and soap multiple times does not clean the rust away in its entirety.

I recommend three additional steps to clean highly stained sea glass. To clean serious embedded dirt, sand, and seaweed:

Try scrubbing your sea glass clean with a toothbrush and dish soap.

Or put your dirty pieces into a dishwasher baby bottle basket. These are plastic, dishwasher safe baskets I was able to find when my twins were little and I needed a receptacle to keep their smaller nursing bottle parts from getting thrown about and lost in the dishwasher. They can be found in the infant and toddler supplies section of the grocery store or a specialty baby shop. Simply load the basket with a handful of your dirty sea glass pieces and run the glass through the next time you run the dishwasher. It is not uncommon to have to run the batch through more than once. Remember, you are washing off years and years of biological material.

A third technique I've had to implement is to soak my stained pieces in a bath solution of a calcium lime rust-removing product. This can take days of soaking, rinsing, re-soaking, and re-rinsing. Take care to follow the directions on the product bottle and use just like you would on anything glass, ceramic, or tile.

There will be a certain point at which your love for your sea glass is not worth the amount of time and energy you may have, by this time, put into the cleaning process. And so your final option is to resolve yourself to leaving your uniquely discolored sea glass pieces as they were found and enjoy the conversation pieces you've got in your collection.

This otherwise beautiful, smooth blue sea glass piece has a nugget of metal solidly stuck to the surface. It has rusted and left a discolored residue around the encrusted metal attachment. I've tried and tried to remove the metal chunk but with no luck.

how to sort, store, and categorize

When most collectors arrived home after any given beach walk or vacation, they'd do just what my family and I do. We'd set our pieces down on a coffee table somewhere. Or we'd leave the pieces in the sandy bottomed pockets of our beach jackets. Sometimes a pile would be released into a candle holder on a bookshelf.

Soon, the cabin or vacation motel would house a museum full of piles, large and small, that represented each individual beach hike. And there were always two or three pieces sitting on top of the dryer along with two pennies and a gum wrapper that had previously gone through the wash.

Many times after sea glass collecting, we will drop our finds piece by piece into one large bag. Upon arriving home, we will dump the bag on a kitchen table or office desk and get to sorting. Sorting is one of my favorite things to do, but after a day of walking along the beach with your neck and back in a stooped position, it is difficult to continue the posture while sitting at a table to tackle a sorting project.

Most collectors have some sort of organization to their collection. I've seen several gatherers rely on plastic Tupperware-style storage trays or a fisherman's tackle box. I like the tackle box idea because the container can be locked tightly closed and carried around by a handle.

I have also observed an elaborate egg carton storage system. This allowed the collector to stack several trays on top of each other. The stacking also helped keep glass from falling out of the compartments because the containers were sandwiched closely together.

Don't be fooled by all the blues. These piles represent about forty years of collecting, piece by piece, one hike at a time. My friend, Teresa, has had these stashed away for years in drawers, dishes, and baggies in her island home. And some of them are as small or even smaller than peppercorns. A good number of pieces are even as miniscule as grains of sand.

At West Coast Sea Glass we categorize our collection by:

1. Ocean or body of water the piece was found along
2. The sea glass condition; whether it's rough and "unfinished" or smooth and flawless
3. Color; see page 133
4. Size and shape of the pieces

Pictured here are several cloth-lined wicker baskets. You will notice that the larger unsegregated baskets hold the larger collections by color like white and brown. And the smaller baskets with sections hold one color per smaller section like pink, red, and yellow. These are the colors of which we don't have such a large quantity.

things to do with your collection

Some collectors put their items into jewelry, candle holders, mosaics, wind catchers, and other art.

We were contracted by a New Jersey interior design firm who was customizing a beach house for a client. The client wanted an eight-foot sea glass, sea shell, and beaded wall hanging created by hand for the bedroom.

It was a fantastical undertaking as we collaborated and ended up creating two of them for two different clients. It took our team of eight about two months to design, measure, collect, drill, tie, and hang what ended up being thousands of pieces.

The piece was so large that it was quite difficult to find a blank wall in our work studio on which to hang the piece so that we could work with it. Good thing my daughter is six feet tall. She became the artist who worked on most of the top strands.

Wall hangings and sun catchers make nice use of your less than perfectly frosted glass or your glass that might have a rough edge or two.

Putting your sea glass in a piece of art is always a nice way to hang onto it and admire it for years. Some sea glass art is considered so well made that it is given as heirloom jewelry or fine art for families to keep for generations.

Imagine finding a special piece on a special hike that marked a momentous day and then having it set in fine jewelry to keep forever.

Here, two sea glass triangles are paired next to one another in this hammered silver cuff bracelet. The cobalt blue piece most likely originates from decades old medicine bottle glass. The emerald green piece comes from more common Pacific Ocean bottle glass.

I recommend Lindsay Furber's book titled *Sea Glass Jewelry* for those of you interested in crafting with your sea glass. In it, Furber takes the reader through step-by-step instructions on how to create simple jewelry and ornamentation with your sea glass.

resources

Sources that were influential in the writing and researching of this book include:

books

Van Rensselaer, Stephen. *Early American Bottles and Flasks* (Borden Pub Co: 1971).

newspapers/magazines

National Geographic
"The Shard Way," August 2007
The Telegraph
"Drowning in Plastic: The Great Pacific Garbage Patch is twice the size of France":
www.telegraph.co.uk/earth/environment/5208645/Drowning-in-plastic-The-Great-
Pacific-Garbage-Patch-is-twice-the-size-of-France.html

websites

Department of Ecology, State of Washington
www.ecy.wa.gov
Food Production Daily
www.foodproductiondaily.com/
House of Glass, Inc.
www.thehouseofglassinc.com/
North American Sea Glass Association
www.seaglassassociation.org/
West Coast Sea Glass
www.WestCoastSeaGlass.com

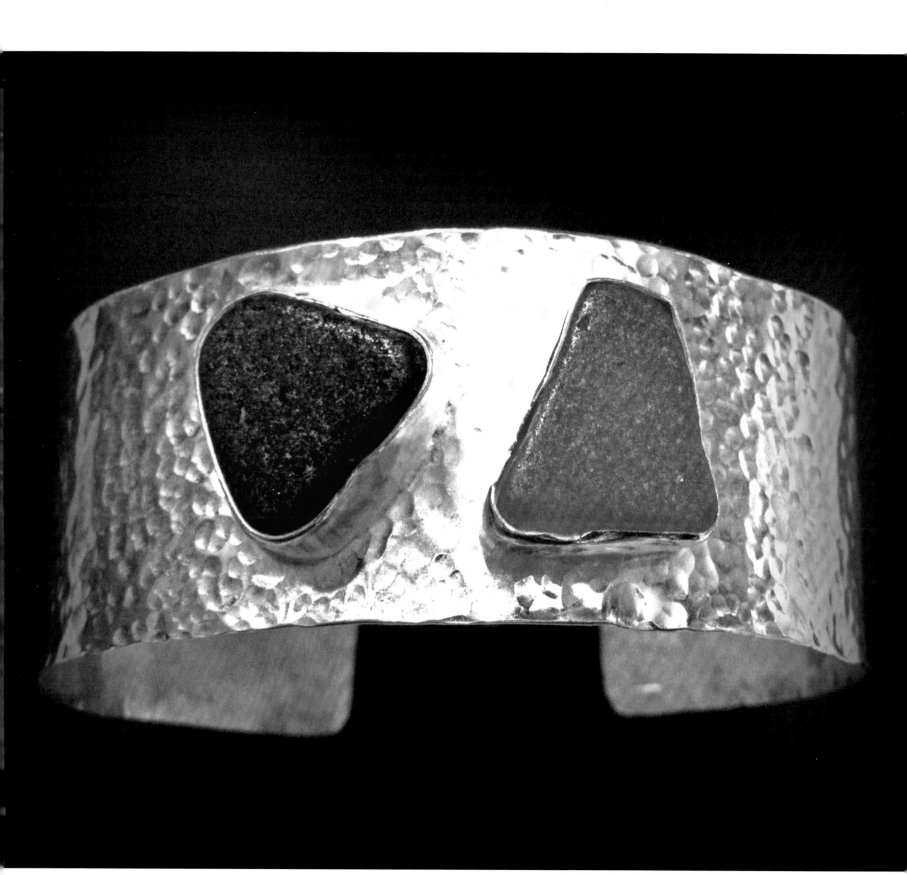

author bio

Mary Beth Beuke is the founder of the International Sea Glass Association, president of the North American Sea Glass Association (from 2005 to 2010), and published sea glass writer, photographer, and speaker. She is a Pacific Northwest native with one of the world's most elaborate sea glass collections. She is owner, artist, and collector for West Coast Sea Glass, and her sea glass art jewelry is now found in more than fifty galleries all over the world. She's journeyed the planet for sea glass for a lifetime and her story and collection have been featured on the Travel Channel, in *National Geographic, Smithsonian, Coastal Living, Parade, Radio New Zealand, Ocean Home, Seattle Times, Sunset,* and more. She exhibits her collection at museums and libraries often. Mary Beth has won Startup Nation's "Leading Moms in Business" award. She lives and uses her sea glass collection in fine art jewelry while working from two studios along the shores in Washington State where she is also the founder of the Smarties, Smart Art Instruction School for artist entrepreneurs. She has three children and enjoys kayaking and public speaking in her spare time.

about west coast sea glass

I have been a beachcomber all of my life and the beaches of the Pacific NW have long been my home. It was along the shores of the Salish Sea that West Coast Sea Glass came into being.

The company's journey started out with my history in jewelry design. I've studied jewelry making, taken metal smith courses and have been creating wearable art for over thirty years. I was educated in graphics reproduction and photography (among other things), and have been drawn to the visual arts and creating art with my hands. As a young adult I supported myself as a semi-professional photographer, owned my own dark room, and also taught photo lab work at the college level. I've traveled much using my camera as my window to the world, won two national photography competitions, and I now use sea glass as a vibrant and colorful photo subject.

I spent more than twenty years as a youth director. I ran outdoor programs, camps, and retreats which took me out on the water; kayaking, managing beach trips, white water rafting, and organizing shoreline cleanups.

My personal and professional years along the ocean's shores presented me with a lifetime of collecting sea glass and studying its conditioning, history, and rarity. I soon learned that what I had was a treasure trove. And then I found myself in a very unique space when it all made sense and where several of my worlds came together. I was a jewelry maker who because of my location and love for the coast, has stumbled upon a vast collection of sea glass.

Lindsay who now works with West Coast Sea Glass also grew up along the shore but near the Washington ferry docks. Her family and mine ended up moving six houses away from each other and we shared the same stretch of beach along the Puget Sound. Though we'd both beachcombed all our lives, it was there that West Coast Sea Glass was born one sunny day. The West Coast Sea Glass line features genuine, handpicked sea glass from rugged Pacific beaches combed all the way from South in Mexico to Alaska's islands and more.

With two individual lifetimes of beachcombing behind us, we combined our finds in the late '90s and West Coast Sea Glass became a business reality. With an extensive and